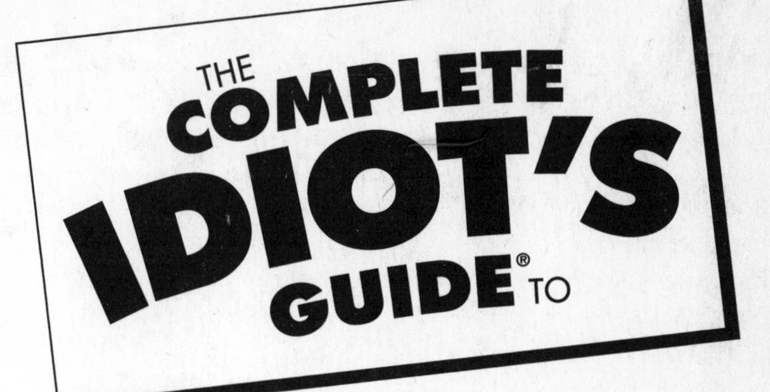

Astronomy

Fourth Edition

Astronomy

Fourth Edition

by Christopher De Pree, Ph.D., and Alan Axelrod, Ph.D.

A member of Penguin Group (USA) Inc.

ALPHA BOOKS

Published by the Penguin Group

Penguin Group (USA) Inc., 375 Hudson Street, New York, New York 10014, USA

Penguin Group (Canada), 90 Eglinton Avenue East, Suite 700, Toronto, Ontario M4P 2Y3, Canada (a division of Pearson Penguin Canada Inc.)

Penguin Books Ltd., 80 Strand, London WC2R 0RL, England

Penguin Ireland, 25 St. Stephen's Green, Dublin 2, Ireland (a division of Penguin Books Ltd.)

Penguin Group (Australia), 250 Camberwell Road, Camberwell, Victoria 3124, Australia (a division of Pearson Australia Group Pty. Ltd.)

Penguin Books India Pvt. Ltd., 11 Community Centre, Panchsheel Park, New Delhi—110 017, India

Penguin Group (NZ), 67 Apollo Drive, Rosedale, North Shore, Auckland 1311, New Zealand (a division of Pearson New Zealand Ltd.)

Penguin Books (South Africa) (Pty.) Ltd., 24 Sturdee Avenue, Rosebank, Johannesburg 2196, South Africa

Penguin Books Ltd., Registered Offices: 80 Strand, London WC2R 0RL, England

International Standard Book Number: 978-1-59257-719-4
Library of Congress Catalog Card Number: 2007937343

10 09 08 8 7 6 5 4 3 2 1

Interpretation of the printing code: The rightmost number of the first series of numbers is the year of the book's printing; the rightmost number of the second series of numbers is the number of the book's printing. For example, a printing code of 08-1 shows that the first printing occurred in 2008.

Printed in the United States of America

Publisher: *Marie Butler-Knight*
Editorial Director/Acquiring Editor: *Mike Sanders*
Managing Editor: *Billy Fields*
Development Editor: *Michael Thomas*
Production Editor: *Kayla Dugger*
Copy Editor: *Nancy Wagner*
Cartoonist: *Shannon Wheeler*
Cover Designer: *Bill Thomas*
Book Designer: *Trina Wurst*
Indexer: *Brad Herriman*
Layout: *Brian Massey*
Proofreader: *Aaron Black*

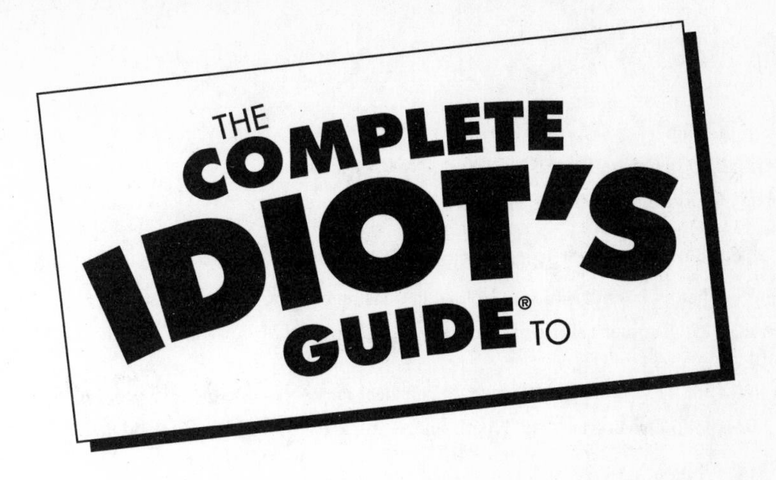

Astronomy

Fourth Edition

by *Christopher De Pree, Ph.D., and Alan Axelrod, Ph.D.*

A member of Penguin Group (USA) Inc.

ALPHA BOOKS

Published by the Penguin Group

Penguin Group (USA) Inc., 375 Hudson Street, New York, New York 10014, USA

Penguin Group (Canada), 90 Eglinton Avenue East, Suite 700, Toronto, Ontario M4P 2Y3, Canada (a division of Pearson Penguin Canada Inc.)

Penguin Books Ltd., 80 Strand, London WC2R 0RL, England

Penguin Ireland, 25 St. Stephen's Green, Dublin 2, Ireland (a division of Penguin Books Ltd.)

Penguin Group (Australia), 250 Camberwell Road, Camberwell, Victoria 3124, Australia (a division of Pearson Australia Group Pty. Ltd.)

Penguin Books India Pvt. Ltd., 11 Community Centre, Panchsheel Park, New Delhi—110 017, India

Penguin Group (NZ), 67 Apollo Drive, Rosedale, North Shore, Auckland 1311, New Zealand (a division of Pearson New Zealand Ltd.)

Penguin Books (South Africa) (Pty.) Ltd., 24 Sturdee Avenue, Rosebank, Johannesburg 2196, South Africa

Penguin Books Ltd., Registered Offices: 80 Strand, London WC2R 0RL, England

International Standard Book Number: 978-1-59257-719-4
Library of Congress Catalog Card Number: 2007937343

10 09 08 8 7 6 5 4 3 2 1

Interpretation of the printing code: The rightmost number of the first series of numbers is the year of the book's printing; the rightmost number of the second series of numbers is the number of the book's printing. For example, a printing code of 08-1 shows that the first printing occurred in 2008.

Printed in the United States of America

Publisher: *Marie Butler-Knight*
Editorial Director/Acquiring Editor: *Mike Sanders*
Managing Editor: *Billy Fields*
Development Editor: *Michael Thomas*
Production Editor: *Kayla Dugger*
Copy Editor: *Nancy Wagner*

Cartoonist: *Shannon Wheeler*
Cover Designer: *Bill Thomas*
Book Designer: *Trina Wurst*
Indexer: *Brad Herriman*
Layout: *Brian Massey*
Proofreader: *Aaron Black*

Contents at a Glance

Contents

Appendixes

Foreword

I was going to be a marine biologist until my parents bought me a telescope when I was in the seventh grade. I took it outside and set it up in my backyard in suburban Fort Worth, Texas. The sky was clear, and the stars were out. One bright star caught my attention. I pointed the small telescope at it to figure out why it was so brilliant. Pointing that little telescope took a bit of work, but I finally centered the bright dot in the finder scope and carefully looked through the main eyepiece. What I saw changed my life forever.

Instead of just a bright speck made brighter by the light-gathering power of the telescope, what appeared was a small, bright, crescent-shaped object. I was floored. I had no idea what I was looking at. It looked kind of like the crescent Moon, but was much smaller and had no surface features. I ran inside to get the guidebook that came with the telescope and within a few minutes had figured out that I was looking at the planet Venus. I ran inside again, got my parents and brother and sister to come outside, and showed them what I had discovered. I am not sure they were as impressed as I was. At least none of them became astronomers. Maybe you have to make the personal effort to learn about the sky to truly get excited about astronomy.

If you are reading this book, then you are about to make that personal step. Inside the pages of *The Complete Idiot's Guide to Astronomy, Fourth Edition*, is all the information you need to slake your thirst for astronomical knowledge. From the solar system to the most distant reaches of our universe, we discuss every kind of object, including what we know about it and how we know what we know, as well as the implications of this knowledge. We present current results in easily understood ways with special additional information set off from the rest of the material. Striking images and pictures from telescopes in space and on the ground—including a wealth of color images on CD for the first time in this edition—show you what you cannot see with your eye, the detailed beauty of the heavens.

If a book like *The Complete Idiot's Guide to Astronomy, Fourth Edition*, had been around when I first began to study astronomy, I probably would have done better in my college classes. This is no joke! Much of the information in this book is cutting-edge stuff; even some researchers might not know some of the information in these pages. Impress your astronomer friends or your regular friends—professional astronomers are pretty rare—at parties, or, if you are a younger reader, your science teacher, by reciting some of the new results you find in this book. Including new and cutting-edge results in a book for novice astronomers is a great thing and a unique value in this volume.

I work in Washington, D.C., advocating for increased spending for basic research, especially in astronomy. I am an astronomical lobbyist. Scientific and technical issues often intimidate members of Congress. "Get me an expert," they often say, "I wasn't trained as a scientist." But they are quite happy to speak at length on social issues, taxation, the economy, international relations, and so on, even if they weren't trained in those fields (most of them are lawyers).

This is a common feeling in our country. Science is somehow thought to be especially difficult or only understood through considerable effort by very smart people. Yet nothing could be further from the truth. Nonscientists can easily understand cutting-edge results, and everyone should know the basics. This book makes the hard stuff easy to understand and the easy stuff easier to understand. You'll see.

I've known one of the authors, Chris De Pree, since he was wearing professional "diapers." We shared an office while working on our doctorate degrees and made home-brewed beer on the weekends (most of the time it tasted good). Aside from his somewhat messy habits, poor taste in music, ability to whistle perfectly out of tune, and small grunting noises he makes when concentrating, he was a good office mate. He is a phenomenal author, and I have to say that being able to write this foreword has been a great honor. Plus, I got a pre-publication copy of the book for free! Chris's editors and co-author have made sure none of his messy habits remain in this volume and that all his creativity and expertise are front-and-center. I am sure you will enjoy reading its pages as much as I have.

—Kevin B. Marvel, Ph.D.

Executive officer, American Astronomical Society

Introduction

You are not alone.

Relax. That statement has nothing to do with the existence of extraterrestrial life—though we do get around to that, too. For the present, it applies only to our mutual interest in astronomy. For we (the authors) and you (the reader) have come together because we are the kind of people who often look up at the sky and have all kinds of questions about it. But this habit hardly brands us as unique. Astronomy, the scientific study of matter in the universe, is among the most ancient of human studies. The very earliest scientific records we have—from Babylon, from Egypt, from China—all concern astronomy.

Recorded history spans about 5,500 years, and the recorded history of astronomy starts at the beginning of that period. Humans have been sky watchers for a very, very long time.

And yet astronomy is also among the most modern of sciences. Although we possess the collected celestial observations of some 50 centuries, almost all that we know about the universe we learned in the twentieth century, and we have gathered an enormous amount of essential knowledge since the development of radio astronomy in the 1950s. In fact, the lifetime of any reader of this book, no matter how young, is filled with astronomical discoveries that merit being called milestones. Indeed, in the three years that separate this fourth edition from the third, astronomers have come to breathtaking new conclusions about the nature and fate of the universe. (If you just can't wait, jump to Chapter 17.) We'd call these new results earthshaking—but, because it's the universe we're talking about, that would be a serious understatement.

Astronomy is an ancient science on the cutting edge. Great discoveries were made centuries ago; great discoveries are being made today. And great leaps forward in astronomical knowledge have often followed leaps forward in technology: the invention of the telescope, the invention of the computer, and the development of fast, cheap computers. So much is being learned every day that we've been asked to bring out a revised edition of this book, the fourth in eight years. And even more recent discoveries will be on the table by the time you read this latest edition.

Yet you don't have to be a government or university scientist with your eager fingers on millions of dollars' worth of equipment to make those discoveries. For if astronomy is both ancient and advanced, it is also universally accessible: up for grabs.

The sky belongs to anyone with eyes, a mind, imagination, a spark of curiosity, and the capacity for wonder. If you also have a few dollars to spend, a good pair of binoculars or a telescope makes more of the sky available to you. (Even if you don't want to spend the money, chances are your local astronomy club will let you use members' equipment if you come and join them for a cold night under the stars.) And if you have a computer, we have a CD. One of the most exciting features of the fourth edition of *The Complete Idiot's Guide to Astronomy* is a collection of 200 of the most spectacular astronomical images ever made. If—as it should—the CD just whets your appetite for more, with an Internet connection you—yes, you—have access to much of the information that the government's investment in people and equipment produces: the very latest images from the world's great telescopes and from a wealth of satellite probes, including the Hubble Space Telescope and the Mars Global Surveyor. This information is all free for the downloading. (See Appendix C for some starting points in your online searches.)

We are not alone. No science is more inclusive than astronomy.

Nor is astronomy strictly a spectator sport. You don't have to peek through a hole in the fence and watch the game. You're welcome to step right up to the plate. Many new comets are discovered by astronomy buffs and backyard sky watchers as well as Ph.D. scientists in domed observatories. Most meteor observations are the work of amateurs. You can even get in on such seemingly esoteric fields as radio astronomy and the search for extraterrestrial intelligence.

We'd enjoy nothing more than to help you get started on your journey into astronomy. Here's a map.

How This Book Is Organized

Part 1, "Eyes, Telescopes, and Light," orients you in the evening sky, and introduces you to the basic ingredients of astronomical observation, telescopes, and the photons of light that they catch.

Part 2, "Worlds Without End," begins with a visit to our nearest neighbor, the Moon, and then ventures out into the rest of the solar system for a close look at the planets and their moons, as well as asteroids and comets.

Part 3, "To the Stars," begins with our own Sun, taking it apart and showing how it works. From our Sun, we venture beyond the solar system to the other stars and learn how they are meaningfully observed. In the end, we explore the very strange realm of neutron stars and black holes.

Part 4, "Way Out of This World," pulls back from individual stars to take in entire galaxies, beginning with our own Milky Way. We learn how astronomers observe, measure, classify, and study galaxies and how those galaxies are all rushing away from us at incredible speed. The section ends with the so-called active galaxies, which emit unimaginably huge quantities of energy and can tell us much about the origin and fate of the universe.

Part 5, "The Big Questions," asks how the universe was born (and offers the Big Bang theory by way of an answer), and then if (and how) the universe will end. Finally, we explore the possibilities of extraterrestrial life and even extraterrestrial civilizations.

At the back of the book, you'll find three appendixes that define key terms, list upcoming eclipses, provide star charts, and suggest sources of additional information, including great astronomy websites.

Extras

In addition to the main text, illustrations, and the CD in this *The Complete Idiot's Guide to Astronomy, Fourth Edition*, you'll also find other types of useful information, including definitions of key terms, important statistics and scientific principles, amazing facts, and special subjects of interest to sky watchers. Look for these features:

Astronomer's Notebook

This feature highlights important statistics, scientific laws and principles, measurements, and mathematical formulas.

Astro Byte

Here you'll find startling astronomical facts and amazing trivia. Strange—but true!

def•i•ni•tion

These boxes define some key terms used in astronomy.

Close Encounter

In these boxes, you'll find discussions elaborating on important events, projects, issues, or persons in astronomy.

Acknowledgments

I would like to acknowledge a few people who have made becoming and being an astronomer much more fun. Dr. Jon Kolena taught me as an undergraduate physics major at Duke University when I was a senior. I had never considered a career in astronomy before taking his challenging, engaging class. Dr. Wayne Christiansen is a professor of astronomy at the University of North Carolina at Chapel Hill. He sent me as a green first-year graduate student to a summer institute at the National Radio Astronomy Observatory (NRAO) in Socorro, New Mexico, in 1991, and I thank him for that. And my office mate as a graduate student in Socorro was my friend and colleague Kevin Marvel, now with the American Astronomical Society. His humor, enthusiasm, friendship, and intelligence have been an inspiration.

I also would like to acknowledge the employees of the National Radio Astronomy Observatory. From my days in Socorro as a graduate student to my return trips as a professor to Socorro, New Mexico; Green Bank, West Virginia; and Charlottesville, Virginia; this dedicated group of astronomers, engineers, staff, and support personnel have been like an extended family.

Finally, thanks to Alan for making a dream of mine, writing science for a popular audience, come true.

—Chris De Pree

My thanks to Chris, a great teacher, brilliant astronomer, and wonderful co-author, and to my family, Anita and Ian, for hurtling through space with me.

—Alan Axelrod

Trademarks

Part 1

Eyes, Telescopes, and Light

You surely have looked up at the sky before. Humans always have. Maybe you can find the Big Dipper and even Orion—or at least his Belt—but, for the most part, all the stars look pretty much the same to you, and you can't tell a star from a planet.

The first chapter of this part takes a look at the constellations; the second introduces the telescope and some basic ideas about light as a wave; and the third moves through the electromagnetic spectrum and beyond visible light.

Chapter 1

Naked Sky, Naked Eye: Finding Your Way in the Dark

In This Chapter

- Observations with the naked eye
- The celestial sphere
- Orienting yourself among the stars
- Celestial coordinates and altazimuth coordinates
- Identifying constellations

From the time our earliest ancestors first looked up into the heavens and tried to figure out what it all meant, the night sky has always been our companion. In our world today, the stars we often study are the human variety, as we wonder how much they made in their last movie and who is still with whom. But long ago, humans looked to the heavens for their stories. They looked for patterns that illustrated myths and legends, and as they looked, they seemed to see the patterns they sought.

This chapter tells you all they saw.

Sun Days

We've become jaded—and a bit spoiled—by the increasingly elaborate and costly special effects producers incorporate into today's sci-fi flicks, but none of us these days is nearly as spoiled as the sky most of us look at.

Imagine yourself as one of your ancestors, living 10,000 years ago. Your reality consists of a few tools, some household utensils, perhaps buildings (the city-states were beginning to appear along the Tigris River), and, of course, all that nature has to offer: trees, hills, plants, rivers, streams—and the sky.

The sky is the biggest, greatest, most spectacular object you know. During the day, a brightly glowing disk from which all light and warmth emanate crosses the sky. Announced in the predawn hours by a pink glow on the eastern horizon, the great disk rises, then arcs across the sky, deepening toward twilight into a ruddy hue before slipping below the horizon to the west. Lacking electric power, your working hours are largely dictated by the presence of the Sun's light.

Flat Earth, Big Bowl

As the Sun's glow fades and your eyes become accustomed to the night, the sky gradually fills with stars. Thousands of them shimmer blue, silvery white, some gold, some reddish, seemingly set into a great dark bowl, the celestial sphere, overarching the flat Earth on which you stand.

But wait. Did we say there were thousands of stars in the night sky?

Maybe that number has brought you back through a starlit 10,000 years and into the incandescent lamplight of your living room or wherever you are reading this as you think: "But *I've* never seen thousands of stars!"

As we said earlier, from many locations, our sky is spoiled. The sad fact is that, nowadays, fewer and fewer of us can see anything like the 6,000 or so stars that *should* be visible to the naked eye on a clear evening. Ten thousand years ago, the night sky wasn't lit up with the light pollution of so many sources of artificial illumination that we have today. Unless you travel far from city lights, in our modern world, you might go through your entire life without *really* seeing the night sky.

Man in the Moon

Even in our smog- and light-polluted skies, however, the Moon shines bright and clear. Unlike the Sun, which appears uniform, the surface of the Moon has details we can see, even without a telescope. Even now, almost four decades after human beings walked, skipped, drove, and even hit a golf ball across the lunar surface, the Moon holds wonder. Bathed in its silver glow, we might feel a connection with our ancestors of 10 millennia ago. Like them, we see in the lunar blotches the face of the "Man in the Moon."

If the face of the Moon presented a puzzle to our ancestors, the way the Moon apparently changed shape surely also fascinated them. One night, the Moon might be invisible (a "new moon"); then, night by night, it would appear to grow (wax), becoming a crescent; and one week later, it would be a first quarter moon (which is a half moon in shape).

Through the following week, the Moon would continue to wax, entering its *gibbous* phase, in which more than half of the lunar disk was seen. Finally, about two weeks after the new moon, all of the lunar disk would be visible as the full moon would rise majestically at sunset. Then, through the next two weeks, the Moon would appear to shrink (wane) night after night, passing back through the waning gibbous, third quarter, and waning crescent phases, until it became again the all-but-invisible new moon.

def•i•ni•tion

> **Gibbous** is a word from Middle English that means "bulging"—an apt description of the Moon's shape between its quarter phase and full phase.

The cycle takes a little more than 29 days—a month, give or take a day—and it should be no surprise that the word "month" is derived from the word "moon." In fact, just as our ancestors learned to tell the time of day from the position of the Sun, so they measured what we call weeks and months by the lunar phases. The lunar calendar is of particular importance in many world religions, including Judaism and Islam. For those who came before us, the sky was more than something to marvel at. The ancients became remarkably adept at using the heavens as a great clock and calendar. Nature was not kind, though, in giving us units of time. The day, month, and year are not evenly divisible into one another, and there are 365.25 days (set by the Earth's rotation) in a year (set by the Earth's orbit around the Sun), a fact that has caused much consternation to calendar makers over the centuries.

Neil Armstrong took this picture of fellow astronaut "Buzz" Aldrin, about to join him on the surface of the Moon, July 20, 1969.

(Image from arttoday.com)

Lights and Wanderers

Early cultures noticed that the bowl above them rotated from east to west. They concluded that what they saw rotating was the celestial sphere—which contained the stars—and not the individual stars. All the stars, they noticed, moved together; their positions relative to one another remained unchanged. (That the stars "move" because of Earth's rotation was an idea that would be thousands of years in the making.)

The coordinated movement of the stars was in dramatic contrast to something else the ancient sky watchers noticed. Although the vast majority of stars were clearly fixed in the rotating celestial sphere, a few—the ancients counted five—seemed to meander independently, yet regularly, across the celestial sphere. The Greeks called these five objects *planetes*, "wanderers," and, like nonconformists in an otherwise orderly society, the wanderers would eventually cause trouble.

Celestial Coordinates

Later, you'll find out why we no longer believe that the celestial sphere represents reality; however, the notion of such a fixed structure holding the stars is still a useful model for modern astronomers. It helps us communicate with others about the positions of the objects in the sky. We can orient our gaze into the heavens by thinking of the point of sky directly above Earth's North Pole as the north celestial pole and the point below the South Pole as the south celestial pole. Just as Earth's equator lies midway between the North and South Poles, so the celestial equator lies equidistant between the north and south celestial poles. Think of it this way: if you were standing at the North Pole, then the north celestial pole would be directly overhead. If you were standing at the equator, the north and south celestial poles would be on opposite horizons. And if you were standing at the South Pole, the south celestial pole would be directly overhead.

Astronomers have extended to the celestial sphere the same system of latitude and longitude that describes earthly coordinates. The lines of latitude, as you might recall from geography, run parallel with the equator and measure angular distance north or south of the equator. On the celestial sphere, *declination* corresponds to latitude and measures the angular distance above or below the celestial equator. Celestial declination is expressed in degrees + (above) or – (below) the celestial equator. The star named Betelgeuse, for example, is at a declination of +7 degrees, 24 minutes (1 minute = 1⁄60 degree).

In the latitudes of the United States, stars directly overhead have declinations in the +30 to +40 degree range. On a globe, the lines of longitude run vertically from pole to pole. They mark angular distance measured east and west of the so-called prime meridian (that is, 0 degrees), which by convention and history has been fixed at Greenwich Observatory in Greenwich, England. On the celestial sphere, *right ascension* (R.A.) corresponds to longitude. Although declination is measured as an angle (degrees, minutes, and seconds), right ascension is measured in hours, minutes, and seconds, increasing from west to east, starting at 0. This zero point is taken to be the position of the Sun in the sky on the first day of spring (the vernal equinox). Because Earth rotates once approximately every 24 hours, the same objects will return to their positions in the sky approximately 24 hours later. After 24 hours, Earth has rotated through 360 degrees, so that each hour of R.A. corresponds to 15 degrees on the sky.

If the celestial poles and the celestial equator are projections of earthly coordinates (the poles and the equator), why not simply imagine right ascension as projections of lines of longitude?

There is a good reason why we don't. Think of it this way: the stars in the sky above your head in winter are different from those in summer. For example, in the winter we see the constellation Orion, but in the summer, Orion is gone, hidden in the glare of a much closer star, the Sun. Now, although the stars above you are changing daily, your longitude (in Chicago, for example) is not changing. So the coordinates of the stars cannot be fixed to the coordinates on the surface of Earth. As we'll see later, this difference comes from the fact that in addition to spinning on its axis, Earth is also orbiting the Sun.

Measuring the Sky

The true value of the celestial coordinate system is that it gives the absolute coordinates of an object so that two observers anywhere on Earth can direct their gaze to the exact same star. When you want to meet a friend in the big city, you don't tell her that you'll get together "somewhere downtown." You give precise coordinates: "Let's meet at the corner of State Street and Madison Street." Similarly, the right ascension and declination shows precisely where in the sky to look.

The celestial coordinate system can be confusing for the beginning sky watcher. However, an understanding of this system of coordinates can help the novice observer locate the North Star and know approximately where to look for planets.

A simpler way to measure the location of an object in the sky as observed from your location at a particular time involves two angles, *azimuth* and *altitude.* You can use angles to divide up the horizon by thinking of yourself as standing at the center of a circle that's flat on the ground around you. A circle can be divided into 360 degrees (and a degree can be subdivided into 60 minutes, and a minute sliced into 60 seconds). When you decide which direction is 0 degrees (the convention is due north), you can measure, in degrees, precisely how far an object is from that point. Now after you have taken care of your horizontal direction, you can fix your vertical point of view by imagining an upright half circle extending from horizon to horizon. Divide this circle into 180 degrees, with the 90-degree point directly overhead. Astronomers call this overhead point the zenith.

def•i•ni•tion

Altitude (angular distance above the horizon) and **azimuth** (compass direction expressed in angular measure east of due north) are **altazimuth coordinates.**

Altitude and azimuth are the coordinates that, together, make up the *altazimuth coordinate* system, and, for most people, they are quite a bit easier to use than celestial coordinates of declination and right ascension. An object's altitude is its angular distance above the horizon, and its compass direction, called azimuth, is measured in degrees increasing clockwise from due north. Thus east is at 90 degrees, south at 180 degrees, and west at 270 degrees.

Altazimuth coordinates, while perhaps more intuitive than the celestial coordinate system, do have a serious shortcoming. They are valid only for your location on Earth at a particular time of day or night. In contrast, the celestial coordinate system is universal because it moves with the stars in the sky. For this reason, star catalogs list the right ascension and declination of objects, not their altitude and azimuth coordinates, which are always changing!

Degrees of Separation

In *The Kids in the Hall*, one character would look at people far away through one eye and pretend to crush their heads between his thumb and forefinger. If you try this trick yourself, you'll notice that people have to be at least five or so feet away from you for their heads to be small enough to crush. Those heads don't actually get smaller, of course, just the angular size of the heads does. As things get more distant, they appear smaller—that is, their angular size is reduced.

The celestial sphere is an imaginary construct, and we often don't know the exact distances between us and the objects we see. Fortunately, to locate objects in the sky, we don't need to know their distances from us. We get that information in other ways, which we will discuss later. Now, from our perspective on Earth, two stars might appear to be separated by the width of a finger held at arm's length when they are actually many trillions of miles distant from each other. You could try to fix the measurement between two stars with a ruler, but where would you hold the measuring stick? Astronomers use concepts called angular size and angular separation to discuss the apparent size in the sky or apparent distance between two objects in the sky. For example, if two objects were on opposite horizons, they would be 180 degrees apart. If one were on the horizon and the other directly overhead, they would be 90 degrees apart. You get the picture. Also, a degree is made up of even smaller increments. One degree is made up of 60 minutes (or arcminutes), and a minute is divided into 60 seconds (arcseconds).

Let's establish a quick and dirty scale. The full moon has an angular size of about half a degree or 30 arcminutes or 1,800 arcseconds (which are all equivalent). Now that you know the full moon is about half a degree across, you can use its diameter to gauge other angular sizes.

You can use your hand to estimate angles greater than a half-degree. Look at the sky. Now extend your arm in front of you with your wrist bent so that the back of your hand is facing you. Spread your thumb, index finger, and pinky fully, and fold your middle finger and ring finger down so you can't see them. Voilá—you have a handy measuring device! The distance from the tip of your thumb to the tip of your index finger is about 20 degrees (depending on the length of your fingers!). From the tip of your index finger to the tip of your pinky is 15 degrees; and the gap between the base of your index finger and the base of your pinky is about 10 degrees.

Celestial Portraits

You now have some rough tools for measuring separations and sizes in the sky, but you still need a way to anchor your measurements, which, remember, are relative to where you happen to be standing on Earth. We need the celestial equivalent of landmarks.

We know human brains are natural pattern makers as we have all seen elephants and lions masquerading as clouds in the sky. Present the mind with the spectacle of thousands of randomly placed points of light against a sable sky, and soon it will start "seeing" some pretty incredible pictures. Fortunately for us, our ancestors had vivid imaginations, and so the constellations—arbitrary formations of stars that are perceived as figures or designs—became such pictures, many of them named after mythological heroes, whose images (in the western world) the Greeks created by connecting the dots.

Astronomer's Notebook

Of the 88 constellations, 28 are in the northern sky and 48 are in the southern sky. The remaining dozen lie along the ecliptic—the circle that describes the path that the Sun takes in the course of a year against the background stars. This apparent motion is actually due to Earth moving around the Sun. These 12 constellations are the zodiac, familiar to many as the basis of the tradition of astrology. All but the southernmost 18 of the 88 constellations are at least sometimes visible from part of the United States.

By the second century C.E., Ptolemy listed 48 constellations in his *Almagest*, a compendium of astronomical knowledge. Centuries later, during the late Renaissance, more constellations were added, and a total of 88 are internationally recognized today. We really cannot say the constellations were discovered, because they don't exist except in the minds of those who see them. Grouping stars into constellations is an arbitrary act of the imagination and to present-day astronomers is merely a convenience. In much the same way that states are divided into counties, the night sky is divided into constellations. The stars thus grouped generally have no physical relationship to one another. Nor do they necessarily even lie at the same distance from Earth; some are much farther from us than others. So, remember, we simply imagine, for the sake of convenience, that they are embedded in the celestial sphere.

If the constellations are outmoded figments of the imagination, why bother with them?

The answer is that they are convenient (not to mention poetic) celestial landmarks. "Take a right at Hank's gas station," you might tell a friend. What's so special about that particular gas station? Nothing—until you endow it with significance as a landmark. Nor was there anything special about a group of physically unrelated stars—until someone endowed them with significance. Now these constellations can help us find our way in the sky, and to the casual night observer, they can be more useful than either the celestial or altazimuth coordinate system.

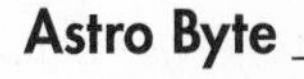

Astro Byte

An abundance of stars retain their Arabic names, a testament to the many contributions of Arabian astronomers: Aldebaran, Mizar, Alcor, and Betelgeuse are a few of these stars.

Enjoy the constellations. The pleasures of getting to know them can occupy a lifetime, and pointing them out to your friends and family can be fun. Nevertheless, recognizing them as the products of human fantasy and not the design of the universe, modern astronomy has only limited use for them. But still we hear the echoes of mythology in modern discoveries.

The Least You Need to Know

- For the ancients, even without telescopes, the night sky was a source of great fascination, which we can share.
- To view the sky meaningfully, you need a system for orienting yourself and identifying certain key features. Celestial coordinates and altazimuth coordinates are two such systems.
- Astronomers use angular size and angular distance to describe the apparent sizes and separations of objects in the sky.
- Constellations are imaginative groupings of stars perceived as images, many influenced by Greek mythology; however, these groupings are arbitrary, reflecting human imagination rather than any actual relationships between those stars.
- Constellations are useful as celestial landmarks to help orient your observations.

Chapter 2

Collecting Light

In This Chapter

- Light as energy that conveys information
- A look at the spectrum and waves
- An introduction to the telescope
- The Hubble Space Telescope and other cutting-edge projects
- Astronomy: getting in on the action

You have every reason and right to consider the night sky the greatest free show in the universe. The great beauty of the sky strongly attracts most amateur stargazers, and in fact, most professional astronomers. But the sky is more than beautiful. Celestial objects are full of information just waiting to be interpreted—information like how distant the stars and galaxies are, how large they are, and whether they are moving toward or away from us.

How does the information reach us? It travels to us in the form of electromagnetic radiation, a small fraction of which is visible light. In this chapter, we begin by defining light and then explain how we can most effectively collect it to see more of the solar system and the universe.

Slices of Light

The universe is ruled by the tyranny of distance. That is, the universe is so vast that we are able to see many things we will never be able to visit. Light travels at extraordinary speeds (about 984,000,000 feet—300,000,000 meters—every second), but the light that we now see from many objects in the sky left those sources thousands, millions, or even billions of years ago. It is possible, for example, to see the Andromeda galaxy (over 2 million *light-years* away), even with the naked eye, but it's highly unlikely humans will ever travel there.

def•i•ni•tion

A **light-year** is the distance light travels in one year—about 6 trillion miles (9,461,000,000,000,000 meters). In the vastness of space beyond the solar system, astronomers use the light-year as a basic unit of distance.

We can't travel at anywhere near the speed of light. Right now, the fastest rockets are capable of achieving 30,000 miles per hour (48,000 km/h), or 262,980,000 miles per year (423,134,820 km/y). Maybe—someday—technology will enable us at least to approach the speed of light, but that still means a trip of 2 million years up and 2 million back.

Why not go faster than the speed of light? According to our understanding of space and time, the speed of light is an absolute speed limit that cannot be exceeded.

So revel in the fact that, on an ordinary night, you are able to gaze at the Andromeda galaxy, an object so distant that no human being will ever visit it.

Spaceships might be severely limited as to how fast they can travel, but the information conveyed by electromagnetic radiation can travel at the speed of light. In fact, the photons we receive from Andromeda left that galaxy long before *Homo sapiens* walked the Earth. But everything we know about Andromeda and almost all other celestial bodies, we know by analyzing their electromagnetic radiation.

Electromagnetic radiation transfers energy and information from one place to another, even in the vacuum of space. The energy is carried in the form of rapidly fluctuating electric and magnetic fields and includes visible light in addition to radio, infrared, ultraviolet, x-ray, and gamma-ray radiation.

The type of energy and information created and conveyed by electromagnetic radiation is more complex than that created and conveyed by the waves generated by a splash in the water.

Making Waves

Electromagnetic radiation sounds like dangerous stuff—and, in fact, some of it is. But the word *radiation* need not set off air raid sirens in your head. It just describes the way energy is transmitted from one place to another without the need for a direct physical connection between them. In this book, we use it as a general term to describe any form of light. It is important that radiation can travel without any physical connection, because space is essentially a vacuum; that is, much of it is empty (at least as far as any scale meaningful to human beings goes). If you went on a space walk clicking a pair of castanets, no one would hear your little concert. Sound is transmitted in waves, but not as radiation. Unlike electromagnetic radiation, sound waves require some medium (such as air) to travel in.

The "electromagnetic" part of the phrase denotes the fact that the energy is conveyed in the form of fluctuating electric and magnetic fields. These fields require no medium to support or sustain them.

Anatomy of a Wave

We can understand how electromagnetic radiation is transmitted through space if we appreciate that it involves waves. What is a wave? The first image that probably jumps to mind is that of ocean waves. And ocean waves do have some aspects in common with the kind of waves that we use to describe electromagnetic radiation. One way to think of a wave is that it is a means by which energy is transmitted from one place to another without physical matter being moved from place to place. Or you might think of a wave as a disturbance that carries energy and that occurs in a distinctive and repeating pattern. A rowboat out in the ocean will move up and down in a regular way as waves pass it. The waves do transmit energy to the shore (think of beach erosion), but the rowboat will stay put.

Astronomer's Notebook

Wave frequency is expressed in a unit of wave cycles per second, called hertz, abbreviated Hz. Wavelength and frequency are inversely related; that is, if you double the wavelength, you automatically halve the frequency, and if you double the frequency, you automatically halve the wavelength. Multiply wavelength by frequency, and you get the wave's velocity. For electromagnetic radiation, wavelength multiplied by frequency is always c, the speed of light.

That regular up-and-down motion that the rowboat experiences is called simple harmonic motion.

Waves come in various shapes, but they all have a common anatomy. They have *crests* and *troughs*, which are, respectively, the high points above and the low points below the level of an undisturbed state (for example, calm water). The distance from crest to crest (or trough to trough) is called the *wavelength* of the wave. The height of the wave—that is, the distance from the level of the undisturbed state to the crest of the wave—is its *amplitude.* The amount of time it takes for a wave to repeat itself at any point in space is its *period.* On a pond, the period is the time between the passage of wave crests as seen by an observer in the bobbing rowboat. The number of wave crests that pass a given point during a given unit of time is called the *frequency* of the wave. If many crests pass a point in a short period of time, we have a high-frequency wave. If few pass that point in the same amount of time, we have a low-frequency wave. The frequency and wavelength of a wave are inversely proportional to one another, meaning that as one gets bigger, the other gets smaller. *High*-frequency radiation has *short* wavelengths.

The parts of a wave.

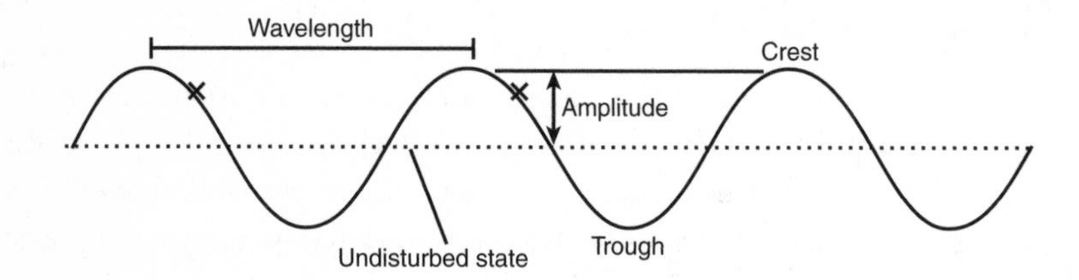

New Wave

If space, as we have said, is very nearly a vacuum, how do waves move through it?

This is a question that vexed physicists for centuries, and at first, most scientists refused to believe that space could be empty. They knew it didn't have air, as on Earth, but they suggested it was filled with another substance, which they called "the ether." But a series of experiments in the late nineteenth century made it clear that ether didn't exist and that although light could be studied as a wave, it was a very different kind of wave than, say, sound.

Big News from Little Places

When the Greek philosopher Democritus (ca. 460–370 B.C.E.) theorized that matter consists of tiny particles he called "atoms," he was partially right. But the story doesn't end there. Atoms can be further broken down into electrons, protons, and neutrons, and the latter two are made of even smaller things called quarks. Electrons carry a negative electric charge, and protons carry a positive charge. Neutrons have a mass almost equal to a proton, but as their name implies, neutrons are neutral, with no positive or negative charge. All charged particles (like protons and electrons) are surrounded by what we call an electric field; those in motion produce electromagnetic radiation.

The British physicist James Clerk Maxwell (1831–1879) first explored what would happen if such a charged particle were to oscillate, or move quickly back and forth. It created a disturbance that traveled through space—*without the need for any medium!*

Particles in space are getting banged around all the time. Atoms collide; electrons are accelerated by magnetic fields; and each time they move, they pull their fields along with them, sending electromagnetic "ripples" out into space.

In short, this changing electric and magnetic field transmits information about the particle's motion through space. A field is not a substance, but a way in which forces can be transmitted over great distances without any physical connection between the places.

Let's turn to a specific example: a star is made up of innumerable atoms, most of which are broken into innumerable charged particles due to unimaginably hot stellar temperatures. A star produces a great deal of energy, which causes particles to be in constant motion. In motion, the charged particles are the center points of electromagnetic waves (disturbances in the electromagnetic field) that move off in all directions. A small fraction of these waves reaches the surface of Earth, where they encounter other charged particles. Protons and electrons in your eyes, for instance, oscillate in response to the fluctuations in the electric field. As a result, you perceive light: an image of the star.

Astronomer's Notebook

All electromagnetic waves—whether visible light, invisible radio waves, x-rays, or gamma radiation—move (in a vacuum) at the speed of light, approximately 186,000 miles per second (299,274 km/s). That's fast, but it's hardly an infinite, unlimited speed. Remember the Andromeda galaxy? We can see it, but the photons of light we just received from the galaxy are 2 million years old. Now *that's* a long commute!

If you happened to have, say, the right kind of infrared-detecting equipment with you, electrons in that equipment would respond to a different wavelength of vibrations originating from the same star. Similarly, if you were equipped with sufficiently sensitive radio equipment, you might pick up a response to yet another set of proton and electron vibrations.

Remember, it is not that the star's electrons and protons have traveled to Earth, but that the waves they generated so far away have excited other electrons and protons here.

You might think of the light from your reading lamp as being very different from the x-rays your dentist uses to diagnose an ailing tooth, but both are types of electromagnetic waves, and the only difference between them is their wavelengths. Frequency and wavelength of a wave are inversely proportional to one another, meaning that if one of them gets bigger, the other one must get smaller. The particular wavelength produced by a given energy source (a star's photosphere—its visible outer layer—or a planetary atmosphere) determines whether the electromagnetic radiation produced by that source is detected at radio, infrared, visible, ultraviolet, x-ray, or gamma-ray wavelengths.

The waves that produce what we perceive as visible light have wavelengths of between 400 and 700 nanometers (a nanometer is 0.000000001 meter) and frequencies of somewhat less than 1015 Hz. Light waves, like the other forms of electromagnetic radiation, are produced by the change in the energy state of an atom or molecule. These waves, in turn, transmit energy from one place in the universe to another. The special nerves in the retinas of our eyes, the emulsion on photographic film, and the pixels of a CCD (Charge Coupled Device) electronic detector are all stimulated (energized) by the energy transmitted by waves of what we call visible light. That is why we—and these other devices—"see."

def•i•ni•tion

The **electromagnetic spectrum** is the complete range of electromagnetic radiation, from radio waves to gamma waves and everything in between, including visible light.

The outer layers of a star, its photosphere, consist of extremely hot gas radiating some fraction of the huge amounts of energy that a star generates in its core through nuclear fusion. That energy is emitted at some level in all portions of the *electromagnetic spectrum*, so that when you look at a distant or nearby star (the Sun) with your eyes, you are receiving a small portion of that energy.

Buckets of Light

Of course, the fraction of the emitted energy you receive from a very distant star—or even a whole galaxy, like far-off Andromeda—is very small, diminished by the square of the distance (but never reaching zero). Imagine a sphere surrounding a distant star. As the sphere becomes larger and larger (that is, as you get farther and farther from the star), the same amount of energy passes through ever larger spheres. Your eye (or your telescope) can be thought of as a very tiny fraction of the sphere centered on that distant star. You are collecting only as much light from the distant source as falls into your "light bucket." If your eye is a tiny "bucket," then a 4-inch amateur telescope is a slightly larger one, and the Hubble Space Telescope is even larger. The larger the bucket, the more light you can "collect." And if you collect more light in your bucket, you get more information.

One early question among astronomers (and others) was, "How can we build a better bucket than the two little ones we have in our head?" The answer came in the early seventeenth century.

The Telescope Is Born

In 1608, Dutch lens makers discovered that if they mounted one lens at either end of a tube and adjusted the distance between the lenses, the lens at one end would magnify an image focused by the lens at the other end of the tube. In effect, the lens at the far end of the tube gathered and concentrated (focused) more light energy than the eye could focus on its own. The lens near the eye enlarged that concentrated image to various degrees. This world-changing invention was dubbed a telescope, a word from Greek roots meaning "far-seeing."

Close Encounter

Using binoculars or a telescope, you might be able to see Venus in its phases, from thin crescent to full. During much of the year, the planet is bright enough to see even in daylight. Venus is closer to the Sun than we are, and that fact keeps it close to the Sun in the sky (never more than 47 degrees, or about a quarter of the sky from horizon to horizon). The best times to observe Venus are at twilight, just before the sky becomes dark, or just before dawn. The planet will be full when it is on the far side of the Sun from us and crescent when it is on the same side of the Sun as we are. With a telescope on a dark night you might be able to observe the "ashen light" phenomenon: when the planet is at quarter phase or less, a faint glow makes the unilluminated face of Venus visible.

Many, perhaps most, inventions take time to gain acceptance. Typically, there is a lapse of more than a few years between the invention and its practical application.

But this was not so with the telescope. By 1609, within a year after the first telescopes appeared, the Italian astronomer Galileo Galilei (1564–1642) demonstrated their naval utility and was soon using them to explore the heavens as well. The largest of his instruments was quite small, with only modest magnifying power, but Galileo was able to use this tool to describe the valleys and mountains on the Moon, to observe the phases of Venus, and to identify the four largest moons of Jupiter.

Refraction ...

Galileo's instrument, like all of the earliest telescopes, was a refracting telescope, which uses a glass lens to focus the incoming light. For all practical purposes, astronomical objects are so far away from us that we can consider that light rays come to us parallel to one another—that is, unfocused. Refraction is the bending of these parallel rays. The convex (bowed outward) piece of glass we call a lens bends the incoming rays so that they all converge at a point called the *focus*, which is behind the lens directly along its axis. The distance from the cross-sectional center of the lens to the focus is called the focal length of the lens. Positioned behind the focus is the eyepiece lens, which magnifies the focused image for the viewer's eye.

Modern refracting telescopes consist of more than two simple lenses. At both ends of the telescope tube, compound (multiple) lenses are used, consisting of assemblies of individual lenses (called elements) designed to correct for various distortions simple lenses produce.

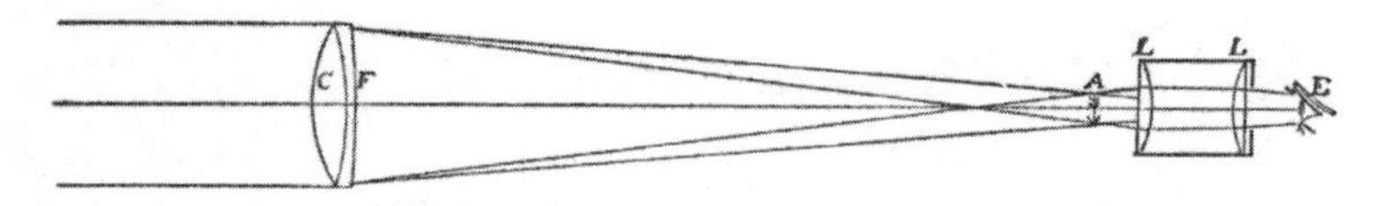

Diagram of a refracting telescope. CF represents the objective lens and LL the eyepiece. The observer's eye is identified by E.

(Image from arttoday.com)

... or Reflection?

The refracting telescope was one of humankind's great inventions, rendered even greater by the presence of a genius like Galileo to use it. However, the limitations of the refracting telescope soon became apparent:

- Even the most exquisitely crafted lens produces distortion.
- Excellent lenses are expensive to produce because both sides of a lens must be precision crafted and polished.
- Generally, the larger the lens, the greater the magnification and the brighter the image; however, large lenses get heavy fast. Lenses have volume, and the potential for imperfections (such as bubbles in the glass) is higher in a large lens. All of this means that large lenses are much more difficult and expensive to produce than small ones.

Recognizing the deficiencies of the refracting telescope, Isaac Newton developed a new design, the reflecting telescope, in 1668.

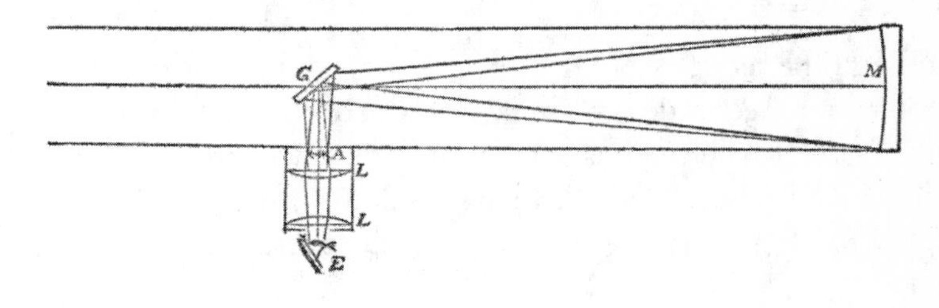

Diagram of a Newtonian reflector. Light enters at the left and is focused by the primary mirror (M) at the back of the telescope. The focused image is sent by a secondary mirror (G) through the eyepiece (LL). The observer's eye is labeled E.

(Image from arttoday.com)

Astro Byte

Newton usually gets sole credit for inventing the reflecting telescope, but another Englishman, John Gregory, actually beat him to it, with a design created in 1663. It was, however, the Newtonian reflector that caught on. The French lens maker Guillaume Cassegrain introduced another variation on the reflector design in 1672. In his design, there is a primary and a secondary mirror, and the focal point of the primary mirror is located behind the primary mirror surface, requiring a hole in the center of the primary mirror.

Instead of the convex lens of a refractor, the reflector uses a concave mirror (shaped like a shallow bowl) to gather, reflect, and focus incoming light. The hollow side of your breakfast spoon is essentially a concave mirror (the other side is a convex one). This curvature means that the focal point is in front of the mirror—between the mirror and the object being viewed. Newton recognized that this was at best inconvenient—your own head could block what you are looking at—so he introduced a secondary mirror to deflect the light path at a 90-degree angle to an eyepiece mounted on the side of the telescope.

Refracting telescope design continued to develop, culminating in the 40-inch (that's the diameter of the principal lens) instrument at Yerkes Observatory in Williams Bay, Wisconsin, installed in 1897. But due to the limitations just mentioned, the biggest, most powerful telescopes have all been reflectors, such as the 100-inch reflector installed at the Mount Wilson Observatory (near Pasadena, California) early in the twentieth century.

Variations on an Optical Theme

Although the refractor and the reflector are the two major types of optical telescopes, there are many variations in reflector design. Some larger reflecting telescopes employ a Cassegrain focus, in which the image from the primary mirror is reflected to a secondary mirror, which again reflects the light rays down through an aperture (hole) in the primary mirror to an eyepiece at the back of the telescope instead of at the side of the telescope.

A coudé-focus (coudé is French for "bent") reflector sends light rays from the primary mirror to a secondary mirror, much like a Cassegrain. However, instead of focusing the light behind the primary mirror, another mirror is employed to direct the light away from the telescope, through an aperture, and into a separate room, called the coudé-focus room. Here astronomers can house special imaging equipment that might be too heavy or cumbersome to actually mount to the barrel of the telescope.

Reflecting telescopes have their problems as well. The presence of a secondary mirror (or a detector, in the case of a prime-focus reflector) means that some fraction of the incoming light is necessarily blocked. Also, the spherical shape of the reflector introduces spherical aberration, light being focused at different distances when reflecting from a spherical mirror. If not corrected, this aberration produces blurred images. One common solution to spherical aberration is to use a very thin "correcting" lens at the top of the telescope. This type of telescope is called a Schmidt-Cassegrain and is a popular design for some high-end amateur telescopes.

Astronomer's Notebook

Astronomers speak of the angular resolution of a telescope, which is a measurement of the smallest angle separating two objects that are resolvable as two objects. Generally, Earth's major optical and infrared telescopes, located at the best sites, can resolve objects separated by as little as 1" (that is, 1 arcsecond, which is $\frac{1}{60}$ of 1 arcminute, which, in turn, is $\frac{1}{60}$ of 1 degree). The theoretical resolution of these telescopes is much higher than this value, but turbulence in Earth's atmosphere means that, except for exceptional nights, this is the best that an Earth-based optical telescope can do. No matter how big the telescope, conventional telescopes cannot have resolutions higher than this value unless they employ "adaptive optics." In recent years, the use of adaptive optics, which uses computer technology to make real-time adjustments in telescope optics to compensate for atmospheric turbulence, has improved the resolution to the subarcsecond level.

Size Matters

In 1948, the Hale telescope at Mount Palomar, California, was dedicated. Its 200-inch (5-meter) mirror was the largest in the world until 1974, when the Soviets completed a 74-ton, 236-inch (6-meter) mirror, which they installed at the Special Astrophysical Observatory in Zelenchukskaya in the Caucasus Mountains.

In 1992, the first of two Keck telescopes, operated jointly by the California Institute of Technology and the University of California, became operational at Mauna Kea, Hawaii. The second Keck telescope was completed in 1996. Each of these instruments combines 36 71-inch (1.8-meter) mirrors into the equivalent of a 393-inch (10-meter) reflector. Not only do these telescopes now have the distinction of being the largest telescopes on Earth, but they are also among the highest (of those based on Earth), nestled on an extinct volcano 2.4 miles above sea level.

Astro Byte

In theory, the 6-meter reflecting telescope in Russia's Caucasus can detect the light from a single candle at a distance of 14,400 miles. However, the presence of Earth's atmosphere and other real-world factors don't permit the practical achievement of this theoretical potential.

The Power to Gather Light

Why do we have this passion for size?

As we mentioned before, the bigger the bucket, the more light you can collect, so the more information you can gather. The observed brightness of an object is directly proportional to the area (yes, area; not diameter) of the primary mirror. Thus, a mirror of 78-inch (2-meter) diameter yields an image 4 times brighter than a 39-inch (1-meter) mirror, because area is proportional to diameter squared, and the square of 2 (2 times 2) is 4. A 197-inch (5-meter) mirror would yield images 25 times brighter (5 times 5) than a 1-meter mirror, and a 393-inch (10-meter) mirror would yield an image 100 times brighter than a 1-meter mirror.

Now, things that are farther away are always going to be fainter. A 100-watt light bulb will appear more faint if it is 1 mile away versus 1 foot away. A telescope that can see fainter objects is able to see things that are farther away; therefore, the bigger the telescope, the more distant the objects that we can view.

The Power to Resolve an Image

Collecting more light is only one advantage of a large telescope. Such instruments also have greater resolving power—that is, the ability to form distinct and separate images of objects that are close together. Low resolution produces a blurred image; high resolution produces a sharp image.

Twinkle, Twinkle

Theoretically, the giant Hale telescope at Mount Palomar is capable of a spectacular angular resolution of .02" (or 20 milliarcseconds); however, because of real-world complications—mostly the presence of the Earth's atmosphere—its practical resolution is about 1" (1 arcsecond). The source of this limit is related to the reason why stars twinkle. Earth's turbulent atmosphere stands between the telescope's gigantic primary mirror and the stars, smearing the image just as it sometimes causes starlight viewed with the naked eye to shimmer and twinkle. If you took a still photograph of a twinkling star through a large telescope, you would see not a pinpoint image, but one that had been smeared over a minute circle of about 1 arcsecond. This smeary circle is called the seeing disk, and astronomers call the effect of atmospheric turbulence *seeing*.

Computer Assistance

Beginning in the late nineteenth century, most serious telescope viewing was done photographically, not in real time, but by studying photographic plates exposed at the focus of a telescope. Photographic methods allowed astronomers to make longer observations, seeing many more faint details than they could ever distinguish with visual observing. In recent years, chemical-based photography has yielded to digital photography, which records images not on film but on CCDs (charge-coupled devices)—in principle the same device at the focal plane of your camcorder lens or digital camera.

CCDs are much more sensitive than photographic film, which means they can record fainter objects in briefer exposure times; moreover, the image produced is digital and can be directly transferred to a computer. Remember the sound of old-fashioned 12-inch vinyl LP records? Even the best of them had a hiss audible during quiet musical passages. CDs, recorded digitally, changed all that by electronically filtering out the nonmusical noise found at high frequencies. Analogous digital computer techniques can filter out the "visual noise" in an image to improve its quality.

Close Encounter

This might be a good time to take a break and turn to your computer. Log on to the World Wide Web and point your browser to the site of the Space Telescope Science Institute at www.stsci.edu and peruse some of the wonderful images from the Hubble Space Telescope (which we will discuss in just a moment).

Fun House Mirrors

Despite atmospheric turbulence, recent technology has made it possible to break the 1-arcsecond barrier.

Adaptive optics systems enable astronomers to correct the distortions introduced by the atmosphere with distortions of their own. The distortions are made to another reflective surface inserted into the optical path, the path that light follows through the telescope. The idea is that if the distortions can be removed quickly enough (in real time), then large telescopes would have both of the advantages that they should have, namely more sensitivity and more resolution.

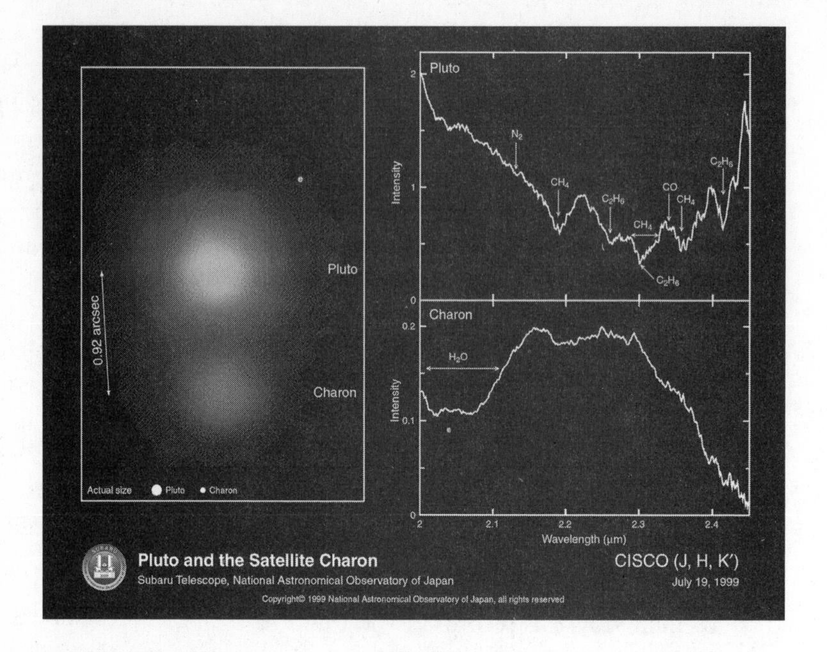

An image of Pluto and its companion Charon taken with the Subaru 8.3-m telescope resolves the two closely orbiting bodies into separate points of light. The apparent separation of Pluto and Charon is only about 0.9", beyond detection without adaptive optics.

An Observatory in Space: The Hubble Space Telescope

Another way to escape the bad seeing caused by Earth's atmosphere is to get above the atmosphere. That is just what NASA, in conjunction with the European Space Agency, did with the Hubble Space Telescope (HST). High above Earth's atmosphere, the HST regularly achieves its theoretical resolution of about less than 0.05" (50 milliarcseconds).

The HST was deployed from the cargo bay of the space shuttle Discovery in 1990. The telescope is equipped with a 94-inch (2.4-meter) reflecting telescope, capable of 10 times the angular resolution of the best Earth-based telescopes and approximately 30 times more sensitive to light, not because it is bigger than telescopes on Earth, but because it is above Earth's atmosphere. Unfortunately, due to a manufacturing flaw, the curvature of the 2.4-meter mirror was off by literally less than a hair (it was too flat by 1/50 of the width of a human hair), which changed its focal length. The telescope still focused light, but not where it needed to, in the plane of the various detectors. Astronauts aboard the shuttle Endeavour rendezvoused with the HST in space in 1993 and

Astro Byte

At a cost of $3 billion, the Hubble Space Telescope is one of the most expensive scientific instruments ever made. However, considering the wealth of scientific information it has produced, the Hubble is a bargain.

made repairs—primarily installing a system of small corrective mirrors. HST then began to transmit the spectacular images that scientists had hoped for and the world marveled at.

The HST was replaced by the Next Generation Space Telescope (NGST), which made way for the James Webb Telescope.

Participatory Astronomy

The cost of nearly $3 billion for the Hubble Space Telescope proves astronomy can be a dauntingly expensive pursuit. Fortunately, you don't have to spend quite that much to get started. In fact, you don't have to spend anything. You can do a lot of observation with the naked eye. Also, many local communities have active amateur astronomers who would be happy to let you gaze at the heavens through their telescopes. Investing in a good pair of binoculars is another good way to start your sky viewing.

Buying—or making—and using your own telescope can be rewarding and even thrilling. It can also be disappointing. For example, you won't be very happy if you can never get away from the light pollution of the city or suburbs. And a cheaply made telescope will frustrate you with dark and blurry images, whereas a more costly device might send you into the black hole of credit card debt or cause a different kind of frustration as you try to deal with an instrument more complicated and temperamental than you really want to deal with.

Take time to learn about amateur telescopes, including what features to look for, how much to spend, and what to do with the thing once you've bought it. Check out any of these books: *Backyard Astronomy: Your Guide to Starhopping and Exploring the Universe*, edited by Robert Burnham (Time Life, 2001); *Turn Left at Orion*, by Guy J. Consolmagno (Cambridge University Press, 1995); and *The Backyard Astronomer's Guide*, by Terence Dickinson and Alan Dyer (Firefly, 2002).

And remember, you can get deeply involved in astronomy without owning or even using a telescope. If you have a personal computer (you do, don't you?), consider purchasing an astronomy software package, such as *The Sky* (from Bisque Software) or *Starry Night* ("Pro" or "Backyard" versions, from Space Software). If you don't want to buy software in shrink wrap, take a look at "CVC's Astronomy Freeware and Software" at www.cvc.org/astronomy/freeware.htm and "Astronomy—Free Software" at http://freeware.intrastar.net/astronmy.htm for a rundown of popular packages, including many that are freely downloadable.

But you don't need to buy or even download free software if you don't want to. Try surfing the Internet into outer space. A great place to start is www.nasa.gov, which has fantastic images of space exploration. Hubble Space Telescope images are available at http://hubble.nasa.gov and elsewhere on the Internet, and images from the Very Large Array and the Green Bank Telescope are available at www.nrao.edu. Another great source is the Jet Propulsion Laboratories web site at www.jpl.nasa.gov. And these are just for starters. Any good search engine will find thrilling images, including many from ongoing space missions as well as earthbound observatories.

The Least You Need to Know

- Light is a form of electromagnetic radiation. Radiation carries energy and conveys information.
- Objects in space produce or reflect the various forms of electromagnetic radiation (including radio, infrared, visible light, ultraviolet, x-rays, and gamma rays); this radiation is what we see with our eyes or detect with special instruments.
- The two basic optical telescope types are the lens-based refractors and the mirror-based reflectors.
- The two main functions of telescopes are to collect light and resolve objects. Larger telescopes (barring the effects of Earth's atmosphere) are better able to perform both functions.
- New technologies, such as adaptive optics, allow ground-based telescopes to achieve much sharper images while maintaining the convenience and lower cost of being on the ground.

Chapter 3

Over the Rainbow

In This Chapter

- Understanding electromagnetic radiation
- Transmission of energy by waves
- Our atmosphere as ceiling and skylight
- Using the electromagnetic spectrum to get information from the sky
- Radiation across the spectrum

The light we receive from distant sources is generated on the tiniest of scales. To explore the largest objects, such as galaxies, we first have to understand the smallest of objects: atoms and the particles that make up atoms. The photons that we detect with our eyes and catch with our telescopes were generated in many different ways: sometimes by electrons hopping between different orbital levels in an atom or other times by the energetic collisions of atomic nuclei. In this chapter, we explore the ways in which photons of light arise, how they get from there to here, and what they can tell us about the objects that we observe.

We have concentrated thus far on optical photons (the ones you can see with your eyes). As it turns out, our eyes respond to "visible" wavelengths because that is where the peak of the emission from the Sun is located in the electromagnetic spectrum. If your eyes were more sensitive to infrared radiation, for example, you would see some things you can't now see (body heat, for one), but you'd miss a lot of other useful stuff, such as the visual appearance of the body that's giving off heat.

In this chapter, we talk more about visible light and the rest of the electromagnetic spectrum, of which visible light is a tiny subset. Think of it this way: if the electromagnetic spectrum is represented by a piano keyboard, then the visible part of the spectrum is but a single key: one note. In the cosmic concerto, there are many notes, and we want to be able to hear them all.

Full Spectrum

Often, when people get excited, they run around, jump up and down, and shout without making a whole lot of sense. But when atomic particles get excited, they can produce energy that is radiated at a variety of wavelengths. In contrast to the babble of an excited human throng, this electromagnetic radiation can tell you a lot, if you have the instruments to interpret it.

One such instrument, the human eye, can interpret electromagnetic radiation in the 400 to 700 *nanometer* (or 4,000 to 7,000 *Angstrom*) wavelength range. A nanometer (abbreviated nm) is one billionth of a meter, or 10^{-9} meter. An Angstrom (abbreviated A) is 10 times smaller, or 10^{-10} meter. But that is only a small part of the spectrum. What about the rest?

The Long and the Short of It

When most people hear the word *radio*, they think of the box in their car that receives radio signals broadcast from local towers, amplifies them, then uses them to drive a speaker, producing sound waves heard with the ears. But radio waves themselves are as silent as optical light or x-rays or gamma rays. They are simply a form of electromagnetic radiation that has very long wavelengths.

So the only difference between radio waves and visible light waves is the length of the wave or the frequency, which is always inversely related to wavelength. Indeed, all forms of electromagnetic radiation represented across what we call the electromagnetic spectrum—radio waves, infrared, visible light, ultraviolet, x-rays, and gamma rays—are transmitted at the speed of light as waves differentiated only by their wavelength and frequency. Radio waves are at the low end of the spectrum, which means their waves are big (on the scale of millimeters to meters in size) and their frequency, therefore, low—in the megahertz (million) to gigahertz (billion) range. Gamma rays, in contrast, are at the high end of the spectrum, with very short wavelengths and very high frequencies.

Close Encounter

Radio waves, visible light, gamma rays—the only difference between these rays is wavelength. But it is quite a difference. Some of the radio waves received by the Very Large Array radio telescope in New Mexico have wavelengths as large as a yard-stick, whereas gamma rays have wavelengths about the size of an atomic nucleus.

The energy of a particular wavelength of electromagnetic radiation is directly proportional to its frequency. Thus, photons of light that have high frequencies and short wavelengths (such as x-rays and gamma rays) are the most energetic, and photons that have low frequencies (radio waves) are the least energetic. Do you ever wonder why the origin stories of comic-book superheroes often involve gamma rays? Those gamma-ray photons carry a lot of energy, and apparently it takes a lot of energy to make a superhero.

Because we are mere mortals, however, we are highly fortunate that Earth's atmosphere absorbs most of the high-energy photons that strike it. Energetic photons tend to scramble genetic material, and the human race wouldn't last long without the protective blanket of ozone in our upper atmosphere. If a massive star were to explode somewhere near Earth in the future, the most harmful effect would be the high-energy photons that would cause a wave of mutations in the next generation.

What Makes Color?

Visible light, we have said, is defined as light with wavelengths from 400 to 700 nm in length, about the dimensions of an average-sized bacterium.

All colors are contained within this tiny range of wavelengths. Just as wavelength (or frequency) determines whether electromagnetic radiation is visible light or x-rays or something else, so it determines what color we see. Our eyes respond differently to electromagnetic waves of different wavelengths. Red light, at the low-frequency end of the visible spectrum, has a wavelength of about 7.0×10^{-7} meters (and a frequency of 4.3×10^{14} Hz). Violet light, at the high-frequency end of the visible spectrum, has a wavelength of 4.0×10^{-7} meters (and a frequency of 7.5×10^{14} Hz). All other colors fall between these extremes, in the familiar order of the rainbow: orange just above red; yellow above orange; then green, blue, indigo, and violet. An easy way to remember this is to think of the name "Roy G. Biv," with every letter in the name representing the first name of a color (red, orange, yellow, green, blue, indigo, and violet).

So-called white light is a combination of all the colors of the visible spectrum. Light is different from paint. Dump together a lot of different colors of paint, and you'll end up with a brownish gray. However, dump together all of the colors of the rainbow, and you'll get "white light." When sunlight is refracted (or bent) as it passes through water droplets in the air, different colors of light are bent to different degrees. The "divided light" then reflects on the back side of the raindrop, and you see the result of this process occurring in a myriad of water droplets as a spectacular rainbow.

Astro Byte

For 25 years, cosmic gamma ray bursts (GRBs) have been one of the great mysteries of modern astronomy. GRBs have given us several clues as to what they might be. First discovered by Earth-monitoring satellites looking for secret nuclear test explosions, GRBs were seen to occur frequently and appeared to be spread evenly over the sky. Their distribution on the sky indicated that they were the result of events happening either very close or very far (in other galaxies). The events were not, for example, seen to be concentrated in the plane of our Galaxy.

Concentrated study and follow-up observations have shown that GRBs appear to arise in distant galaxies. If they are distant and very bright, then the source of the GRB must be a very energetic event. One possible explanation of GRBs is that they are the result of the merger of two neutron stars, the dense remnant cores of exhausted massive stars. This explanation has accounted for many of the known characteristics of GRBs.

Heavenly Scoop

We see celestial objects because they produce energy, and that energy is transmitted to us in the form of electromagnetic radiation. As you will see in later chapters, different physical processes produce different wavelengths (energies) of light. Thus the portion of the spectrum from which we receive light itself is an important piece of information.

Atmospheric Ceilings ...

The information—the news—we get from space is censored by the several layers of Earth's atmosphere. In effect, a ceiling pierced by two skylights surrounds Earth. A rather broad range of radio waves readily penetrates our atmosphere, as does a portion of infrared and most visible light, in addition to a small portion of ultraviolet. Astronomers speak of the atmosphere's *radio window* and *optical window*, which allow passage of electromagnetic radiation of these types. To the rest of the spectrum—lower-frequency radio waves, some lower-frequency infrared, and (fortunately for our survival) most of the energetic ultraviolet rays, x-rays, and gamma rays—the atmosphere is opaque, an impenetrable ceiling. An atmosphere opaque to these wavelengths but transparent to visible light and some infrared is a big reason why life can survive at all on Earth.

... and Skylights

For astronomers, however, there is a downside to the selective opacity of Earth's atmosphere. Observations of ultraviolet, x-ray, and gamma-ray radiation cannot be made from the surface of Earth, but must be made by means of satellites, which are placed in orbit well above the atmosphere.

Until well into the twentieth century, astronomers had no way to "see" most of the nonvisible electromagnetic radiation that reached Earth from the universe. Then along came radio astronomy, which got its start in 1931–1932 and was cranking into high gear by the end of the 1950s. Over the past 40 years or so, much of our current knowledge of the universe has come about through radio observations.

Dark Doesn't Mean You Can't See

On a clear night far from urban light pollution, the sky is indeed dazzling. Just remember that the electromagnetic information your eyes are taking in, wondrous as it is, comes from a very thin slice of the entire spectrum. Earth's atmosphere allows only visible light and a bit of infrared and ultraviolet radiation to pass through a so-called optical window, but it allows a broad portion of the radio spectrum to pass through a radio window.

Anatomy of a Radio Telescope

In principle, a radio telescope works like an optical telescope. It is a "bucket" that collects radio frequency waves rather than visible light waves and focuses them on a detector or receiver. A large metal dish—like a giant TV satellite dish—is supported on a moveable mount. A detector, called a receiver horn, is mounted on legs above the dish (prime focus) or below the surface of the dish (Cassegrain focus). The telescope is pointed toward the radio source, and its huge dish collects the radio waves and focuses them on the receiver, which amplifies the signal and sends it to a computer.

Because the radio spectrum is so broad, astronomers have to decide which portion of the radio spectrum to observe. So they use different receivers for observations at different frequencies. Receivers are either swapped in and out, or (more typically) the radio signal is directed to the correct receiver by moving a secondary reflecting surface (like the secondary mirror in an optical telescope).

Bigger Is Better: The Green Bank Telescope

In the case of radio telescopes, size really does matter. The resolution of a telescope depends not only on its diameter, but also on the wavelength of the detected radiation (the ratio of wavelength to telescope diameter determines the resolution). Radio waves are big (on the order of centimeters or meters), and the telescopes that detect them are correspondingly huge. Also, the radio signals these instruments detect are very faint, and just as bigger optical telescopes with bigger mirrors collect more light than smaller ones, bigger radio telescopes collect more radio waves and image fainter radio signals than smaller ones.

Astronomer's Notebook

Why do radio telescopes have to be so big? A 300-foot or 100-meter dish might seem excessive. The resolution is actually determined by the ratio of the wavelength being observed to the diameter of the telescope. So optical telescopes (which detect short-wavelength optical photons) can be much smaller than radio telescopes, which are trying to detect long-wavelength radio waves and have the same resolution.

Collecting radio signals is just part of the task, however. You might recall from Chapter 2, that for practical purposes, very good optical telescopes located on Earth's surface can resolve celestial objects to 1" (1 arcsecond—1/60 of 1 arcminute, which, in turn, is 1/60 of 1 degree). The best angular resolution that a very large single-dish radio telescope can achieve is some 10 times coarser than this, about 10", and this, coarse as it is, is possible only with the very largest single dish radio telescopes in the world. The National Radio Astronomy Observatory has recently completed the world's largest fully steerable radio telescope. The 100-m dish of the Green Bank Telescope at Green Bank in Pocahontas County, Virginia, has a best resolution of 14".

The world's largest nonsteerable single-dish radio telescope was built in 1963 at Arecibo, Puerto Rico, and uses a dish 300 meters (984 feet) in diameter sunk into a natural valley. Although its great size makes this the most sensitive radio telescope, the primary surface is nonsteerable—totally immobile—and, therefore, is limited to observing objects that happen to pass roughly overhead (within 20 degrees of zenith) as Earth rotates.

The Arecibo Radio Telescope, Arecibo, Puerto Rico.

(Image from National Astronomy and Ionosphere Center)

Interference Can Be a Good Thing

A way to overcome the low angular resolution due to the size of radio waves is to link together a lot of smaller telescopes so they act like one giant telescope. A radio interferometer is a combination of two or more radio telescopes linked together electronically to form a kind of virtual dish, an array of antennas that acts like one gigantic antenna.

The National Radio Astronomy Observatory (NRAO) maintains and operates the Very Large Array (VLA) interferometer on a vast plain near Socorro, New Mexico, consisting of 27 large dishes arrayed on railroad tracks that are laid out in a Y-shaped pattern. Each arm is 12.4 miles (20 km) long, and the largest distance between two of the antennas is 21.7 miles (35 km). As a result, the VLA has the resolving power—but not the sensitivity—of a radio telescope 21.7 miles across.

For radio astronomers who want something even larger than "very large," there is Very Long Baseline Interferometry (VLBI), which can link radio telescopes in different parts of the world to achieve incredible angular resolutions of a thousandth of an arcsecond (.001") or better. From its offices in Socorro, New Mexico, the NRAO also operates the VLBA (Very Long Baseline Array), which consists of 10 radio dishes scattered over the United States, from Mauna Kea, Hawaii, to St. Croix, U.S. Virgin Islands.

Still not big enough? In 1996, Japanese astronomers launched into Earth orbit a radio telescope to be used in conjunction with the ground-based telescopes in order to achieve the resolution of a telescope larger than Earth itself. For more information, see www.nrao.edu.

What Radio Astronomers "See"

Insomnia is a valuable affliction for optical astronomers who need to make good use of the hours of darkness when the Sun is on the other side of Earth. But the Sun is not a particularly bright radio source and, therefore, won't interfere with observations in the radio spectrum. So radio astronomers (and radio telescopes) can work night and day. The VLA, for example, gathers data (or runs tests) 24 hours a day, 363 days a year. Not only is darkness not required, but you can also make radio observations through a cloud-filled sky.

Radio astronomers can observe objects whose visible light doesn't reach Earth because of obscuration by interstellar dust or simply because they emit little or no visible light. The fantastic objects known as quasars, pulsars, and the regions around black holes—all of which you will encounter later in this book—are often faint or invisible optically but do emit radio waves.

The Rest of the Spectrum

Optical astronomy with the naked eye is at least 5,000 years old and probably much older. Optical telescope astronomy is about 400 years old. Radio astronomy is a youthful discipline at about 70 years, if we date its birth from the work of Karl Jansky, the American telecommunications engineer who pioneered the principles of radio astronomy in the early 1930s. But it has been only since the 1970s that other parts of the electromagnetic spectrum have been regularly explored for the astronomical information they might yield. Each new window thrown open on the cosmos has brought in a fresh breeze and enriched our understanding of the universe.

New Infrared and Ultraviolet Observations

Telescopes need to be specially equipped to detect infrared radiation—the portion of the spectrum just below the red end of visible light. Infrared observatories have applications in almost all areas of astronomy, from the study of star formation, cool stars, and the center of the Milky Way, to active galaxies and the large-scale structure of the universe.

IRAS (the Infrared Astronomy Satellite) was launched in 1983 and sent images back to Earth for many years. Like all infrared detectors, though, the ones on IRAS had to be cooled to low temperatures so that their own heat did not overwhelm the weak signals they were trying to detect. Although the satellite is still in orbit, it has long since run out of coolant and can no longer capture images. The infrared capability of the Hubble Space Telescope provided by its NICMOS (Near-Infrared Camera and Multi-Object Spectograph) yielded spectacular results while in operation. Launched on August 25, 2003, the Spitzer Space Telescope (formerly called the Space Infrared Telescope Facility, or SIRTF) observes the universe at wavelengths between 3 and 180 microns (a micron is a millionth of a meter).

And NASA's James Webb Space Telescope (formerly the Next-Generation Space Telescope—NGST), scheduled for launch in 2011, will be optimized to operate at infrared wavelengths and will be cooled passively by a large solar shield.

Ultraviolet radiation, which begins in the spectrum at frequencies higher than those of visible light, is also being studied with new telescopes. Because our atmosphere blocks all but a small amount of ultraviolet radiation, we must make ultraviolet studies by using high-altitude balloons, rockets, or satellites like the Hopkins Ultraviolet Telescope (HUT) flown in 1990 and 1995 in the payload bay of the space shuttle *Endeavour*.

Chandrasekhar and the X-Ray Revolution

Astronomers can now study electromagnetic radiation at the highest end of the spectrum, but because x-rays and gamma rays cannot penetrate our atmosphere, satellites must do all this work. X-rays are detected from very high-energy sources, such as the remnants of exploded stars (supernova remnants) and jets of material streaming from the centers of galaxies. Work began in earnest in 1978 with the launch of the High-Energy Astronomy Observatory (later dubbed the Einstein Observatory), an x-ray telescope. Germany launched the next x-ray telescope, the Röentgen Satellite (ROSAT), in 1990. Finally, in 1999, the Chandra X-ray Observatory (named for astronomer Subrahmanyah Chandrasekhar), the world's premier x-ray instrument, was launched. It has produced unparalleled high-resolution images of the x-ray universe. The Chandra image of the Crab Nebula, home to a known pulsar, showed never-before-seen details of the environment of an exploded star. For recent images, go to www.chandra.harvard.edu, which is continuously updated.

In 1991, the Compton Gamma Ray Observatory (CGRO) was launched by the space shuttle *Atlantis* and was in operation until June 2003. It revealed unique views of the cosmos, especially in regions where the energies involved are very high: near black holes, at the centers of active galaxies, and near neutron stars. You can check out a mission summary of the CGRO at http://cossc.gsfc.nasa.gov.

The Black-Body Spectrum

As Maxwell first described in the nineteenth century, all objects emit radiation at all times because the charged atomic particles of which they are made are constantly in random motion. As these particles move, they generate electromagnetic waves. Heat an object, and its atomic particles will move more rapidly, thereby emitting more radiation. Cool an object, and the particles will slow down, emitting proportionately less and lower-energy electromagnetic radiation. If we can study the spectrum (that is, the intensity of light from a variety of wavelengths) of the electromagnetic radiation emitted by an object, we can understand much about the source. One of the most important quantities we can determine is its temperature. Fortunately, we don't need to stick a thermometer in a star to see how hot it is. All we have to do is look at its light carefully.

Here's how: all objects emit radiation, but no natural object emits all of its radiation at a single frequency. Typically, the radiation is spread out over a range of frequencies. If we can determine how the intensity (amount or strength) of the radiation emitted by an object is distributed across the spectrum, we can learn a great deal about the object's properties, including its temperature.

def•i•ni•tion

A **black body** is an idealized (imaginary) object that absorbs all radiation that falls on it and perfectly reemits all radiation it absorbs. The spectrum that such an object emits is an idealized mathematical construct called a **black-body curve**, which can serve as a model for measuring the peak intensity of radiation emitted by a real object, such as a star.

Physicists often refer to a *black body*, a theoretical (that is, imaginary) object that absorbs all radiation falling upon it and reemits all the radiation it absorbs. The way in which this reemitted energy is distributed across the range of the spectrum is drawn as a *black-body curve.*

Now, no actual object in the physical world absorbs and radiates in this ideal fashion, but we can use the black-body curve as a model against which the peak intensity of radiation from real objects can be measured. The reason is that the peak of the black-body curve shifts toward higher frequencies (and shorter wavelengths) as an object's temperature increases. Thus, an object or region that is emitting mostly very short wavelength gamma-ray photons must be much hotter than one producing mostly longer wavelength radio waves. If we can determine the wavelengths of the peak of an object's electromagnetic radiation emissions, we can determine its temperature.

Home on the Range

Astronomers measure peak intensity with sophisticated scientific instruments, but we all do this intuitively almost every day. For example, on an electric kitchen range, when the knob for one of the heating elements is turned to *off*, the heating element is black in color. This tells you that it might be safe to touch it. But if you were to turn on the element and hold your hand above it, you would feel heat rising and know it was starting to get hot. If you had infrared vision, you would see the element "glowing" in the infrared. As the element grows hotter, it will eventually glow visibly red, and, even without special infrared vision, you would know it was absolutely a bad idea to touch it, regardless of where the control knob happened to be pointing.

At room temperature, the metal of the heating element is black, but as it heats up, it changes color: from dull red to bright red. If you had a very high-voltage electric range and a sufficiently durable heating element, you could crank up the temperature so that it became even hotter. It would emit most of its electromagnetic radiation at progressively higher frequencies.

Now, an object that omits *most* of its radiation at optical frequencies would be *very* hot. And a kitchen range will never (we hope) reach temperatures of 6,000 K, like the Sun. The red color you see from the range is in the "tail" of its black-body spectrum. Even when hot, it is still emitting most of its radiation in the infrared part of the spectrum.

Read Any Good Spectral Lines Lately?

Using the spectrum and armed with the proper instrumentation, astronomers can accurately read the temperature of even very distant objects in space. And even without sophisticated equipment, you can startle your friends by letting them know that Betelgeuse (a *reddish* star) must have a lower surface temperature than the *yellow* Sun.

Astronomers also use the spectrum to learn even more about distant sources. A spectroscope passes incoming light through a narrow slit and prism, splitting the light into its component colors. Certain processes in atoms and molecules give rise to emission at very particular wavelengths. Using such a device, astronomers can view these individual spectral lines and glean even more information about conditions at the source of the light.

Although ordinary white light simply breaks down into a continuous spectrum—the entire rainbow of hues, from red to violet, shading into one another—light emitted by certain substances produces an emission spectrum with discrete emission lines, which are, in effect, the fingerprint of the substance.

Hydrogen, for example, has four clearly observable spectral lines in the visible part of the spectrum (red, blue-green, violet, and deep violet). The color from these four lines (added together as light) is pinkish. These four spectral lines result from the electron that is bound to the proton in a hydrogen atom jumping between particular energy levels. Many other spectral lines are being emitted; it just so happens that only four of them are in the visible part of the spectrum. In our hydrogen atom example, a negative electron is bound to a positive proton. The electron, while bound to the proton, can only exist in certain specific states or *energy levels.* Think of these energy levels as rungs on a ladder. The electron is either on the first rung, the second rung, or the third rung, and so on; it can't be in between. When the electron moves from a higher energy level to a lower one, it gives off energy in the form of a photon. Because the levels the electron can inhabit are limited, only photons of a few specific frequencies are given off. These particular photons are apparent as bright regions in the spectrum of hydrogen: the element's spectral emission lines.

Depending on how you view an astronomical source, you will see different types of spectra. A black-body source viewed directly will produce a continuous spectrum. But if the photons from the source pass through a foreground cloud of material, the cloud (depending on its composition) will absorb certain energies, and you will see a black-body spectrum with certain portions of the spectrum missing or dark. This is called an absorption spectrum. If a cloud of material absorbs energy and then reemits it in a different direction, you will see the result as emission lines, or bright regions in the spectrum. The clouds of hot gas around young stars produce such emission lines.

The light that reaches us from stars carries a lot of information. The color of the object can tell us its temperature; the wavelength of the light reaching us can tell us about the energies involved; and the presence (or absence) of certain wavelengths in a black-body spectrum can tell us what elements are present in a given source. Who knew we could learn so much without actually going anywhere?

The Least You Need to Know

- The difference between visible light and other electromagnetic waves, say, radio waves, is just a matter of wavelength or frequency.
- Unlike sound waves, light waves can travel through a vacuum—empty space—because these waves are disturbances in the electromagnetic field and require no medium (substance) for transmission.
- One way spectral lines arise is by the specific energies given off when electrons jump between energy levels in an atom or when molecules spin at different rates.
- Astronomers use spectroscopes to read the spectral "fingerprint" produced by the light received from distant objects and thereby determine the chemical makeup of an object.
- A radio telescope typically consists of a large parabolic dish that collects and focuses very weak incoming radio signals on a receiver. The signal is then amplified and processed using electronics and computers.
- In recent years, astronomers have launched instruments into orbit that can detect all segments of the electromagnetic spectrum, from infrared, through visible, and on to ultraviolet, x-rays, and gamma rays. The highest frequency radiation (x-rays and gamma rays) comes from some of the most energetic and exotic objects in the universe.

Part 2
Worlds Without End

This part takes an extended tour of the solar system, beginning with an overview of the solar system. After this, we look in detail at the eight major planets in the solar system, the four terrestrial planets and the four jovian planets—as well as Pluto, once considered a planet and now deemed by astronomers something else.

Then we explore our own companion, the Moon, and the moons and rings of the jovian planets. The final chapter in this part takes us to the newly discovered realm of extrasolar planets, worlds beyond our solar system.

Chapter 4

Solar System Family Snapshot

In This Chapter

- A solar system inventory
- Introduction to the Sun
- Planetary stats
- All about asteroids
- Comets, meteors, meteoroids, and meteorites

A snapshot freezes an instant in time. When we think about our solar system, we usually assume it has always been much as it is now and always will be. But what we know firsthand of the solar system (4,000 years of accumulated knowledge) is only a mere snapshot in comparison to its 4.6 billion-year age. It took humankind millennia to reach the conclusion that our planet is part of a solar system, one of many planets spinning on its axis orbiting the Sun. There were centuries of wrestling with the Earth-centered planetary system first presented by Aristotle, then by Ptolemy, trying to make the expected planetary orbits coincide with actual observation.

Knowledge of the solar system arose in some sense as a side product of the real initial goals: to be able to predict the motion of the planets and stars for the purposes of creating calendars and (in some cases) as a means of fortune telling. However, even the earliest astronomers (of whom we know) wanted to do more than predict planetary motion. In the sixteenth and seventeenth centuries when Copernicus, Galileo, Tycho Brahe, and Kepler finally succeeded in finding out what was "really" going on, it was a momentous time for astronomy and human understanding.

Understanding how the planets move is important, of course, but our understanding of the solar system hardly ends with such knowledge. In the last few decades of the twentieth century and now in the twenty-first, astronomers have learned more about the solar system than in all the 400 years since planetary motions were pretty well nailed down. As this chapter shows, the planetary neighborhood is a very interesting place, and our own world, Earth, is unique among the planets as a home to life.

Neighborhood Stroll

Although it appears that about half of all stars are located in binary systems (two stars), our solar system is centered on a single star, the Sun. Eight planets orbit around the Sun (in order of distance from the Sun): Mercury, Venus, Earth, Mars, Jupiter, Saturn, Uranus, and Neptune. (Pluto, long counted as the ninth planet, was recently downgraded from a full-fledged planet to a mere "dwarf planet." Around some of these planets orbit moons—about 90 at latest count. By the 1990s, astronomers had observed more than 6,000 large asteroids, of which approximately 5,000 have been assigned catalog numbers. (Such an assignment is made as soon as accurate orbital data is recorded.) Most asteroids are rather small; it is estimated that there are 1 million with diameters greater than 1 km (or about 3/5 of a mile). Some, perhaps 250, have diameters of at least 62 miles (100 km), while about 30 have diameters of more than 124 miles (200 km). All of these planets and asteroids are the debris from the formation of the Sun that coalesced slowly through the mutual attraction of gravity.

In addition, the solar system contains a great many comets and billions of smaller, rock-size meteoroids.

Some Points of Interest

The orbits of the planets lie nearly in the same plane, except for Mercury, which deviates from this plane by 7 degrees. Between the orbit of Mars and Jupiter, most of the solar system's asteroids are found in a concentrated area known as the asteroid belt.

The orbits of the planets are not equally spaced, tending (very roughly) to double between adjacent orbits the farther a planet is from the Sun.

To say that the distances between the planets and the Sun are very great is an understatement. Interplanetary distances are so great that it becomes awkward to speak in terms of miles or kilometers. For that reason, astronomers have agreed on something called an astronomical unit (A.U.), which is the average distance between Earth and the Sun—that is, 149,603,500 kilometers or 92,754,170 miles.

Let's use this unit to gauge the size of the solar system. The average distance from the Sun to Pluto, one of the largest dwarf planets, is 40 A.U. (3,710,166,800 miles or almost 6 billion km). At just about a million times the radius of Earth, that's quite a distance. Think of it this way: if Earth were a golf ball, Pluto would be a chickpea about 8 miles away, Jupiter would be a basketball about 1 mile away, and the Sun would go floor-to-ceiling in a 10-foot room and be less than a quarter-mile away. However, compared to, say, the distance from Earth to the nearest star (after the Sun), even Pluto is a near neighbor. Forty A.U. is less than 1/1,000 of a light-year, the distance light travels in one year: almost 6 trillion miles. Alpha Centauri, the nearest star system to our Sun, is about 4.3 light-years from us (more than 25 trillion miles). On our golf ball scale, Alpha Centauri would be about 55,000 miles away. Not even Tiger Woods has a drive like that.

More or Less at the Center of It All

Near the center of the solar system—more accurately, at one focus of the elliptical orbits of the planets—is the Sun. The Sun is not just the middle of the solar system; it is most of the solar system, containing more than 99.9 percent of its matter. Jupiter, the largest planet in the solar system, is over 300 times the mass of Earth, but the Sun is more than a *thousand* times more massive than Jupiter and about 300,000 times more massive than Earth.

Planetary Report Card

Let's make a survey of the planets. Here's what we'll be measuring and comparing in the table that follows:

- **Semi-major axis of orbit.** Planets orbit the Sun, not in perfectly circular paths but in elliptical ones. The semi-major axis of an ellipse is the distance from the center of the ellipse to its farthest point. This distance doesn't exactly correspond to the distance from the Sun to the farthest point of a planet's orbit, because the Sun isn't at the exact center of the ellipse, but at one of the ellipse's two foci. We will express this number in A.U.

- **Sidereal period.** This number expresses the time it takes a planet to complete one orbit around the Sun, usually expressed in Earth years.
- **Mass.** This is the quantity of matter a planet contains. The mass of Earth is 5.977×10^{24} kg. We will assign Earth's mass the value of 1.0 and compare the masses of the other planets to it.
- **Radius.** At the equator, the radius of Earth is slightly less than 3,963 miles (6,400 km). We will assign the radius of Earth a value of 1.0 and compare the radii of other planets to it.
- **Number of known moons.** This number is self-explanatory although it is an ever-changing number for the outer planets. The parenthetical figures for Jupiter and Saturn indicate the number of small satellites that have been discovered since 2000.
- **Average density.** This value is expressed in kilograms of mass per cubic meter. The substance of the inner planets is dense and tightly packed; in the outer planets, the densities are typically lower.

Planet	Semi-Major Axis of Orbit (in A.U.)	Sidereal Period (in Years)	Mass (in Earth Masses)	Radius (in Earth Masses)	Moons*	Density (kg/m³)
Mercury	0.39	0.24	0.055	0.38	0	5,400
Venus	0.72	0.62	0.81	0.95	0	5,200
Earth	1.0	1.0	1.0	1.0	1	5,500
Mars	1.5	1.9	0.11	0.53	2	3,900
Jupiter	5.2	11.9	318	11.2	49 (13)	1,300
Saturn	9.5	29.5	95	9.5	48 (12)	700
Uranus	19.2	84	15	4.0	27	1,200
Neptune	30.1	165	17	3.9	13	1,700

**Number of moons for each planet is based on September 2007 data from NASA.*

The Inner and Outer Circles

Astronomers used to divide the planets into two broad categories—with one planet left over. The four planets (including Earth) closest to the Sun are termed the terrestrial planets. The four farthest from the Sun are the jovian planets. Pluto's new status as a "dwarf planet" means that it does not need to fit into either of these categories.

Snapshot of the Terrestrial Planets

Mercury, Venus, Mars, and Earth are called the terrestrial planets because they all possess certain Earth-like (terrestrial) properties. These include proximity to the Sun (within 1.5 A.U.), relatively closely spaced orbits, relatively small masses, relatively small radii, and high density (rocky and solid-surfaced). Compared to the larger, more distant jovian planets, the terrestrials rotate more slowly, possess weak magnetic fields, lack rings, and have few or no moons.

Snapshot of the Jovian Planets

The jovians are far from the Sun and travel in widely spaced orbits. They are massive planets with large radii, yet are of low density with predominantly gaseous makeup and no solid surface. In contrast to the terrestrial planets, they rotate faster, possess strong magnetic fields, have rings, and are orbited by many large moons.

Serving Up the Leftovers

What's the oldest stuff in your refrigerator (aside from that rubbery celery you bought but never ate)? Leftovers! The same is true in the solar system. The fragmentary leftovers of the formation of the Sun and planets are some of the oldest objects in the solar system. For a long time, few scientists paid much attention to this debris or knew much about it. More recently, however, they have come to realize that many significant clues to the origin and early evolution of the solar system are to be found not in the planets but in the smaller bodies, the planetary moons and solar system debris. For the most part, the planets are very active places. Atmospheres have produced erosion, and internal geological activity has erased ancient surfaces. On Earth, weather, water, and tectonic motion have long since "recycled" the original surface of the planet.

So studying the planets can reveal relatively little about the origins of the solar system. However, on moons and asteroids, atmospheres are sparse or nonexistent, and geological activity is minimal or absent. The result? Many of these bodies have changed little since the solar system was born. They are, in effect, cosmic leftovers.

The Asteroid Belt

Of the more than 6,000 asteroids with regular orbits that astronomers have noted and cataloged, most of them are concentrated in the asteroid belt. So far, every asteroid that has been observed orbits in the same direction as Earth and other planets—except one, whose orbit is retrograde (backward, or contrary to the direction of the planets). Although the asteroids move in the same direction—and pretty much on the same plane—as the planets, the shape of their orbits is different. Many asteroid orbits are more eccentric (the ellipse is more exaggerated, oblong) than those of the planets.

Landing on Eros—The Love Boat

In early 2001, an asteroid-exploring probe orbited and finally crash-landed on the surface of an asteroid named Eros. As it approached the asteroid's surface, it sent back tantalizing close-up images of the surface.

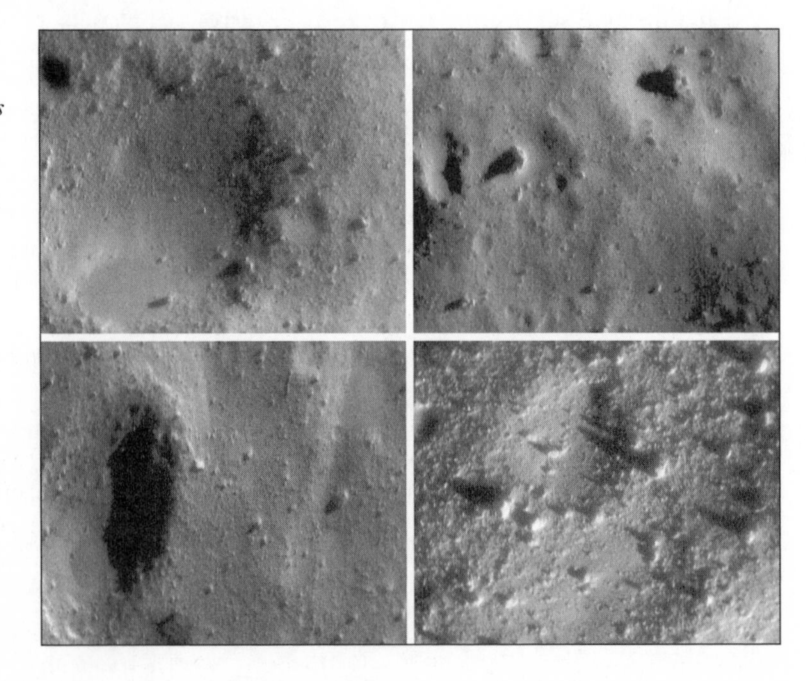

These images of Eros show the closest views of an asteroid ever seen. The top images are 550 m across, and the bottom images are 230 m across. Small boulders dominate the surface.

(Image from NASA)

Rocks and Hard Places

Asteroids are composed of stony as well as metallic—mostly iron—materials and are basically tiny planets without atmospheres. Some asteroids have a good deal of carbon in their composition as well. These, called carbonaceous chondrites, are thought to represent the very first materials that came together to form the objects of the solar system. Carbonaceous chondrites are truly ancient messengers, having avoided change for billions upon billions of years.

Earlier astronomers surmised that asteroids were fragments resulting from various meteoric collisions. Although some of the smaller meteoroids were likely produced this way, we now believe that the major asteroids probably came into being at the time of the formation of the solar system as a whole. Theoretical studies suggest that no planet could have formed at the radius of the asteroid belt (about 3 A.U. from the Sun). The region between Mars and Jupiter is dominated by the gravitational influence of the giant planet Jupiter. This force stirred up the potential planet-forming material, causing it to collide and break up instead of coming together to create a planet-sized object.

The smaller asteroids come in a wide variety of shapes, ranging from nearly spherical, to slab-like, to highly irregular.

Astronomer's Notebook

From 1993 to 1994, the Galileo probe passed through the asteroid belt on its way to Jupiter and took pictures of an asteroid orbited by its own miniature moon. The potato-shaped asteroid named Ida is about 35 miles (56 km) long and is orbited at a distance of roughly 60 miles (97 km) by a rock less than 1 mile in diameter. This little moon is the smallest known natural satellite in the solar system.

Impact? The Earth-Crossing Asteroids

Although most asteroids remain in the asteroid belt, some have highly eccentric orbits that take them out of the belt and across the orbital path of Earth (as well as the paths of other terrestrial planets).

Nearly 100 of these so-called Apollo asteroids have been identified so far, and a number of astronomers passionately advocate funding efforts to identify and track even more because the potential for a doomsday collision with Earth is all too real. With advance warning, some scientists believe, missiles with thermonuclear warheads could be exploded near an incoming asteroid, sufficiently altering its course to avoid Earth or shattering it into a large number of smaller asteroids. Your local movie theater or video store is a good source to study Hollywood's take on these nightmare scenarios, but they are a very credible threat. Project NEAT (Near Earth Asteroid Tracking) is funded by NASA. For more information, see http://neat.jpl.nasa.gov.

It is believed that a few asteroids of more than a half-mile diameter might collide with Earth in the course of a million years. Such impacts would be cataclysmic, each the equivalent of the detonation of several hydrogen bombs. Not only would a great crater, some 8 miles across, be formed, but an Earth-enveloping dust cloud would also darken the skies. Some think the great extinction of the dinosaurs 65 million years ago was due to such an impact. Were the impact to occur in the ocean, tidal waves and massive flooding would result. Earth impacts of smaller objects are not uncommon, but on June 30, 1908, a large object—apparently the icy nucleus of a small comet—fell in the sparsely inhabited Tunguska region of Siberia. The falling object outshone the Sun, and its explosive impact was felt more than 600 miles away. A very wide area of forest was obliterated—quite literally flattened. Pictures from the time show miles of forest with trees stripped and lying on their sides like matchsticks, eerily pointing away from the impact site.

Anatomy of a Comet

The word *comet* derives from the Greek word *kome*, meaning "hair." The name describes the blurry, diaphanous appearance of a comet's long tail.

But the tail is just part of the anatomy of a comet and is not even a permanent part, as it forms only as the comet nears the Sun. For most of its orbit, the comet exists as a main, solid body or the nucleus, which is a relatively small (a few miles in diameter) mass of irregular shape made up of ice and something like soot, consisting of the same elements that we find in asteroids.

The orbit of the typical comet is extremely eccentric (elongated), so that most comets (called long-period comets) travel even beyond Pluto and might take millions of years to complete a single orbit. So-called "short-period" comets don't venture beyond Pluto and, therefore, have much shorter orbital periods.

As a comet approaches the Sun, the dust on its surface becomes hotter, and the ice below the crusty surface of the nucleus sublimates—that is, immediately changes to a gas without first becoming liquid. The gas leaves the comet, carrying with it some of the dust. The gas molecules absorb solar radiation, then reradiate it at another wavelength while the dust acts to scatter the sunlight. This process creates a coma, a spherical envelope of gas and dust (perhaps 60,000 miles across) surrounding the nucleus and a long tail consisting of gases and more dust particles.

A Tale of Two Tails

Most comets actually have two tails. The dust tail is usually broader and more diffuse than the ion tail, which is more linear. The ion tail is made up of ionized atoms—that is, atoms that are electrically charged. Both the dust tail and the ion tail point away from the Sun, but the dust tail is usually seen to have a curved shape that trails the direction of motion of the comet. Careful telescopic or binocular observations of nearby comets can reveal both of these tails.

What we cannot see optically is the vast hydrogen envelope that surrounds the coma and the tail. It is invisible to optical observation.

Common sense tells us that the tail would stream behind the fast-moving nucleus of the comet. This is not the case, however. Both the ion and dust tails point away from the Sun, regardless of the direction of the comet's travel. Indeed, as the comet rounds the Sun and begins to leave its proximity, the tail or tails actually *lead* the nucleus and comet. This is because a comet tail is "blown" like a wind sock by the solar wind, a constant stream of matter and radiation that escapes from the Sun. Astronomers discovered the existence of this solar wind by observing the behavior of comet tails.

"Mommy, Where Do Comets Come From?"

The solar system has two cometary reservoirs, both named after the Dutch astronomers who discovered them. The nearer reservoir is called the Kuiper Belt (after Gerard Peter Kuiper, 1905–1973). The short-period comets, those with orbital periods less than 200 years, are believed to come from this region, which extends from the orbit of Pluto out to several 100 A.U. Comets from this region orbit peacefully unless some gravitational influence sends one into an eccentric orbit that takes it outside of the belt.

Long-period comets, it is believed, originate in the Oort Cloud (after Jan Oort, 1900–1992), a vast area (some 50,000–100,000 A.U. in radius) surrounding the solar system and consisting of comets orbiting in various planes. Oort comets are thought to be distributed in a spherical cloud instead of a disk.

The Oort Cloud is at such a great distance from the Sun that it only extends about one third of the distance to the nearest star. We don't see the vast majority of these comets because their orbital paths, though still bound by the Sun's gravitational pull, never approach the perimeter of the solar system. However, it is believed that the gravitational field of a passing star from time to time deflects a comet out of its orbit within the Oort Cloud, sending it on a path to the inner solar system, perhaps sealing our fate.

After a short-period or long-period comet is kicked out of its Kuiper Belt or Oort Cloud home, it assumes its eccentric orbit indefinitely. It can't go home again. Each time a comet passes close to the Sun, a bit of its mass is boiled away—about 1/1,000 of its mass with each pass. After some 100 passages, a comet typically fragments and continues to orbit as a collection of debris or coalesces with the Sun. As Earth passes through the orbital paths of such debris, we experience meteor showers.

Close Encounter

Comets do not randomly occur, but are regular, orbiting members of the solar system. The most famous comet of all is Halley's Comet, named after the British astronomer Edmund Halley (1656–1742), who published a book in 1705, showing by mathematical calculation that comets observed in 1531, 1607, and 1682 were actually a single comet. Halley predicted the comet would return in 1758, and when it did, it was named in his honor. Subsequent calculations show that Halley's Comet, which appears every 76 years (most recently in 1986), had been seen as early as 240 B.C.E. and was always a source of great wonder and even fear.

Catch a Falling Star

Few astronomical phenomena are more thrilling than the sight of a meteor. Best of all, such sightings are common and require no telescope.

Meteors, Meteoroids, and Meteorites

Meteors are often called shooting stars, although they have nothing at all to do with stars. A meteor is a streak of light in the sky resulting from the intense heating of a narrow channel in Earth's upper atmosphere. The heat generated by friction with air molecules as the meteoroid hurtles through Earth's atmosphere ionizes—strips electrons away from atoms along—a pathway behind this piece of space debris. The ionized path in Earth's atmosphere glows for a brief time, producing the meteor.

Although smaller meteoroids (often called micrometeoroids) are typically the rocky fragments left over from a broken-up comet, the meteor phenomenon is very different from a comet. A meteor sighting is a momentary event. The meteor streaks across a part of the sky, whereas a comet does not streak rapidly and may, in fact, be visible for many months because of its great distance from Earth. A meteor is an event occurring in Earth's upper atmosphere, whereas a comet is typically many A.U. distant from Earth.

Meteor is the term for the sight of the streak of light caused by a meteoroid—which is the term for the actual rocky object that enters the atmosphere. Most meteoroids are completely burned up in our atmosphere, but a few do get through to strike Earth. Any fragments recovered are called *meteorites.*

This meteor crater in Arizona was formed about 50,000 years ago as the result of the impact of an object some 80 feet in diameter. The crater is nearly a mile (over a kilometer) in diameter.

(Image from JPL/NASA)

April Showers (or the Lyrids)

Whenever a comet makes its nearest approach to the Sun, some pieces break off from its nucleus. The larger fragments take up orbits near the parent comet, but some fall behind, so that the comet's path is eventually filled with these tiny micrometeoroids. Periodically, Earth's orbit intersects with a cluster of such micrometeoroids, resulting in a *meteor shower* as the fragments burn up in our upper atmosphere.

def•i•ni•tion

When Earth's orbit intersects the debris that litter the path of a comet, we behold a **meteor shower,** a period when we see more meteors than the average.

Meteor showers associated with certain comets occur with high regularity and are named after the constellation from which their streaks appear to radiate. The following table lists the most common and prominent showers. The shower names are genitive forms of the constellation name; for example, the Perseid shower comes from the direction of the constellation Perseus, the Lyrids from Lyra. The dates listed are those of maximum expected activity, and you can judge the intensity of the shower by the estimated hourly count. The table also lists the parent comet, if known. Consult a monthly astronomy magazine such as *Astronomy* or *Sky & Telescope* to obtain specific peak times for a given year's meteor showers.

Name of Shower	Maximum Activity	Estimated Hourly Count	Parent Comet
Quadrantid	January 3	50	Unknown
Beta Taurid	June 30	25	Encke
Perseid	August 12	50+	1862III (Swift-Tuttle)
Draconid	October 8–9	500+	Giacobini-Zimmer
Orionid	October 20	25	Halley
Leonid	November 16–17	10*	1866I (Tuttle)
Geminid	December 11–17	50–75	3200 Phaeton

**Every 33 years, Earth's orbit intersects the densest part of the Leonid debris path, resulting in the potential for a meteor infall rate of 1,000 a minute! Such an intersection occurred in 1999 and will happen again in 2032.*

The Least You Need to Know

- The eight planets of the solar system are divided into the rocky terrestrial planets (those nearest the Sun: Mercury, Venus, Earth, and Mars) and gaseous jovian planets (those farthest from the Sun: Jupiter, Saturn, Uranus, and Neptune).
- Although the Sun and planets are certainly the major objects in the solar system, astronomers also pay close attention to the minor bodies—asteroids, comets, meteors, and planetary moons—that can tell us a lot about the origin of the solar system.
- Although most asteroids are restricted to highly predictable orbits, a few cross Earth's orbital path, posing a potentially catastrophic hazard.
- Comets and meteor showers present ample opportunities for exciting amateur observation.

Chapter 5

Hard, Rocky Places: The Inner Planets

In This Chapter

- Some vital statistics
- Orbit and rotation
- Atmospheres on the terrestrial planets
- Planetary surfaces
- The geology of the terrestrials
- How everything might have been different

Our two closest neighbors in the solar system, Mars and Venus, are constant reminders of how easily things could have turned out differently here on Earth. Venus is so hot and forbidding that it might be a good place to set Dante's *Inferno*, and although pictures of Mars might resemble the American Southwest, its atmosphere is so cold and thin that it's hardly there at all. Putting a planet a little closer to the Sun or a little farther away can truly make all the difference. Equally amazing is that all three planets fall within what is called the "habitable zone" of the Sun (see Chapter 18), which is the range of distance from the Sun within which water can exist as a liquid on a planet's surface. But only one planet, Earth, has abundant liquid water. Recent imaging of the surface of Mars indicates that water might still exist there in liquid form, albeit only fleetingly.

In this chapter, we take a closer look at the four rocky planets that are closest to the Sun: Mercury, Venus, Mars, and Earth—the terrestrials.

The Terrestrial Roster

Except for Earth, the terrestrial planets, Mercury, Venus, and Mars, are all named after Roman gods. Mercury, the wing-footed messenger of the gods, is an apt name for the planet closest to the Sun; its sidereal period (the time it takes a planet to complete one orbit around the Sun) is a mere 88 Earth days, and its average orbital speed (30 miles per second or 48 km/s) is the fastest of all the planets. Mercury orbits the Sun four times for each Earth orbit.

Named for the Roman goddess of love and fertility, Venus is (to observers on Earth) the brightest of the planets, and, even to the naked eye, quite beautiful to behold. Its atmosphere is not so lovely, however. The planet is completely enveloped by carbon dioxide and thick clouds that consist mostly of sulfuric acid.

The name of the bloody Roman war god, Mars, suits the orange-red face of our nearest planetary neighbor—the planet that has most intrigued observers and that seems, at first glance, the least alien of all our fellow travelers around the Sun.

You looked at some vital statistics of the planets in Chapter 4. Now here are some more numbers, specifically for the terrestrial planets. Notice that the presence of an atmosphere (for Venus and Earth) creates much less variation in surface temperature.

Planet	Mass in Kilograms	Radius in Miles (and km)	Surface Gravity (Relative to Earth)	Rotation Period in Solar Days	Surface Temperature in K
Mercury	3.3×10^{23}	1,488 (2,400)	0.4	59	100–700
Venus	4.9×10^{24}	3,782 (6,100)	0.9	–243*	~730
Earth	6.0×10^{24}	3,968 (6,400)	1.0	1	~290
Mars	6.4×10^{23}	2,108 (3,400)	0.4	1	100–250

**The rotation period is negative because the rotation of Venus is retrograde; that is, the planet rotates on its axis in the opposite direction from the other planets. Viewed from above the North Pole of Earth, all of the planets except Venus rotate counterclockwise. That means that on Venus, the Sun would rise (if you could see it through the thick cloud cover) in the west.*

When we discuss the formation of the solar system in Chapter 8, we'll mention a few observational facts that "constrain" our models of planetary system formation. But as you see, a few rules of planetary motion are immediately apparent. All four terrestrial planets orbit the Sun in the same direction. All except Venus rotate on their axes in the same direction as they orbit the Sun. The orbital paths of the inner four planets are nearly circular. And the planets all orbit the Sun in roughly the same plane.

The solar system is a dynamic and real system, not a theoretical construct, and some interesting exceptions to these rules can give us insight into the formation of this solar system.

Close Encounter

Most of us in the United States are accustomed to the Fahrenheit temperature scale. The rest of the world uses the Celsius (Centigrade) scale. Astronomers, like most scientists, measure temperature on the Kelvin scale. Throughout this book, we have expressed distance in the units familiar to most of our readers: miles (with kilometers or meters given parenthetically). For mass we give all values in kilograms. The Kelvin scale for temperature is conventional and very useful in astronomy; let us explain.

The Fahrenheit scale is really quite arbitrary because its zero point is based on the temperature at which alcohol freezes. What's fundamental about that? Worse, it puts at peculiar points the benchmarks that most of us do care about. For example, water freezes at 32°F at atmospheric pressure and boils at 212°F. The Celsius scale is somewhat less arbitrary because water freezes at 0°C and boils at 100°C at atmospheric pressure. But because the atmospheric pressure of Earth is by no means a universal quantity, astronomers and others looked for more fundamental benchmarks.

The Kelvin scale is least arbitrary of all. It forces us to ask a fundamental question: what is heat?

The atoms and molecules in any matter are in constant random motion, which represents thermal energy. As long as there is atomic or molecular motion, there is heat (even in objects that, to the human senses, feel very cold). We know of no matter in the universe whose atoms and molecules are entirely motionless, but, in theory, such an absolute zero point does exist. The Kelvin scale begins at that theoretical absolute zero, the point at which there is no atomic or molecular motion. On the Fahrenheit scale, that temperature is –459°. On the Celsius scale, it is –273°. On the Kelvin scale, it is merely 0°. Thus, in the Kelvin scale there are no negative temperatures because absolute zero is, well, absolute.

Mercury: The Moon's Twin

In many ways, Mercury has more in common with the lifeless Moon of our own planet than with the other terrestrial planets. Its face is scarred with ancient craters, the result of massive bombardment that occurred early in the solar system's history. These craters remain pristine because Mercury has no water, erosion, or atmosphere to erase them. The closest planet to the Sun—with an average distance of 960,000 miles (1,546,000 km)—Mercury is difficult to observe from Earth and can only be viewed near sunrise or sunset.

Mercury's surface was revealed in detail for the first time in images transmitted by such unmanned probes as *Mariner 10* (in the 1970s). *Mariner 10* also discovered a weak magnetic field. As a result, astronomers concluded that the planet must have a core rich in molten iron. This contention is consistent with the planet's position closest to the center of the solar system, where most of the preplanetary matter—the seed substance that formed the planets—would have been metallic in composition.

This mosaic of Mercury's surface was taken by Mariner 10 *during its approach on March 29, 1974. The spacecraft was about 124,000 miles (200,000 km) above the planet. Note how closely Mercury's surface resembles the Moon's.*

(Image from JPL/NASA)

Lashed to the Sun

In the days before space-based telescopes and probes, earthbound astronomers did the best they could to gauge the rotation of Mercury. The nineteenth-century astronomer Giovanni Schiaparelli observed the movement of what few indistinct surface features he could discern and concluded that, unlike any other planet, Mercury's rotation was synchronous with its orbit around the Sun.

Synchronous orbit means that Mercury always keeps one face toward the Sun and the other away from it, much as the Moon always presents the same face to Earth.

Technology marches on. In 1965, by means of radar imaging, astronomers discovered that Mercury's rotation period was not 88 days, as it was long thought to be, but only 59 days. This discovery implied that Mercury's rotation was not precisely synchronous with its orbit after all but that it rotated three times around its axis for every two orbits of the Sun.

"I Can't Breathe!"

Like Earth's Moon, Mercury possesses insufficient mass to hold—by gravitation—an atmosphere. In the same way that mass attracting mass built up planetesimals—the "embryonic" stage of forming planets, masses no more than several hundred miles across—so the early planets built up atmospheres by hanging on to them with their gravitational pull. If an atmosphere was ever associated with Mercury, the heating of the Sun and the planet's small mass helped it to escape long ago. Without any atmosphere to speak of, the planet is vulnerable to bombardment by meteoroids, x-rays, and ultraviolet radiation, as well as extremes of heat and cold.

Despite the absence of atmosphere, regions at the poles of Mercury remain permanently in shadow, with temperatures as low as 125 K. These regions, and similar regions on Earth's Moon, might have retained some water in the form of ice.

Forecast for Venus: "Hot, Overcast, and Dense"

Venus's thick atmosphere and its proximity to the Sun make for a cruel combination. The planet absorbs more of the Sun's energy and, because of its heavy cloud cover, is unable to radiate away much of the heat. Even before astronomers saw pictures of the planet's surface, they knew it would not be a welcoming place.

Until the advent of radar imaging aboard space probes such as *Pioneer Venus* (in the late 1970s) and *Magellan* (in the mid-1990s), details about the surface of Venus were shrouded in mystery. Optical photons reflect off the upper cloud layers of the planet, and all astronomers can see with even the best optical telescopes is the planet's swirling upper atmosphere. Modern radio imaging techniques (which involve bouncing radio signals off the surface) have revealed a surface of rolling plains punctuated by a pair of raised land masses that resemble Earth's continents. Venus has no coastlines, all of its surface water having long ago evaporated in the ghastly heat. The two land masses, called Ishtar Terra and Aphrodite Terra, are high plateaus in a harsh, waterless world.

The landscape of Venus also sports some low mountains and volcanoes. Volcanic activity on the surface has produced calderas (volcanic craters) and coronae, which are vast, rough, circular areas created by titanic volcanic upswellings of the mantle.

Venus is surely lifeless biologically, but geologically it is very lively. Astronomers think volcanic activity is ongoing, and many believe the significant but fluctuating level of sulfur dioxide above the Venusian cloud cover might be the result of volcanic eruptions. Probes sent to Venus thus far have not detected a *magnetosphere*; however, astronomers believe the planet has an iron-rich core. Scientists reason that the core of Venus might simply rotate too slowly to generate a detectable magnetic field.

def•i•ni•tion

A **magnetosphere** is a zone of electrically charged particles trapped by a planet's magnetic field. The magnetosphere lies above the planet's atmosphere.

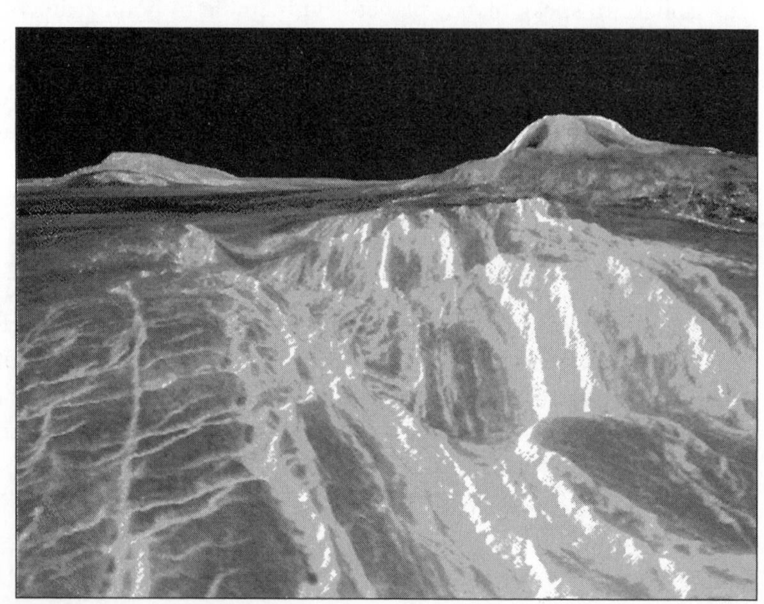

Radar mapping carried out by the Magellan *probe made this image of volcanic domes in the Eistla region of Venus in 1991.*

(Image from JPL/NASA)

The Sun Also Sets (in the East)

As we've seen, Mercury's peculiar rotational pattern can be explained by its proximity to the Sun. But no such gravitational explanation is available for the strange behavior of Venus. If at 59 days, Mercury rotates on its axis slowly, Venus is even more sluggish, consuming 243 Earth days to accomplish a single spin.

What's more, it spins backward! That is, viewed from a perspective above Earth's North Pole, all the planets (terrestrial and jovian) spin counterclockwise—except for Venus, which spins clockwise.

Nobody knows why for sure, but we can surmise that some random event occurred during the formation of the solar system—a collision or close encounter with another planetesimal, perhaps—and caused the planet's rotational oddities. A violent collision, like the one that formed Earth's Moon, might have started Venus on its slow backward spin.

Venusian Atmosphere

Chemically, the atmosphere of Venus consists mostly of carbon dioxide (96.5 percent). The remainder is mostly nitrogen. These are organic gases, which might lead one to jump to the conclusion that life—*some* form of life—might exist on Venus. Indeed, during the 1930s, spectroscopic studies of Venus revealed the temperature of the planet's upper atmosphere to be about 240 K—close to Earth's *surface* temperature of 290 K. Some speculated that the environment of Venus might be a dense jungle.

In the 1950s, for the first time, radio waves penetrated the dense cloud layer that envelops Venus. It turned out that surface temperatures were not 240 K, but closer to 600 K—incompatible with any form of life.

The outlook got only worse from there. Spacecraft probes soon revealed that the dense atmosphere of Venus creates high surface pressure—the crushing equivalent of 90 Earth atmospheres—and that surface temperatures can top 730 K.

And what about those clouds?

On Earth, clouds are composed of water vapor. But Venus shows little sign of water. Its clouds consist of sulfuric acid droplets. Talk about acid rain!

The Earth: Just Right

In our march through the terrestrial planets, the next logical stop would be Earth. We have already mentioned some of the unique aspects of our home planet and will continue to do so through the course of the book. In particular, we will look at Earth as a home to life when we discuss the search for life elsewhere in the Milky Way (Chapter 18).

But let's take a brief moment now to think of Earth as just another one of the terrestrial planets. Earth is almost the same size as Venus and has a rotational period and inclination on its axis almost identical to Mars. How is it, then, that Earth is apparently the only one of these three planets to support life?

As in real estate, it all comes down to three things: location, location, and location. Earth is far enough from the Sun that it has not experienced the runaway greenhouse effect of Venus. It is close enough to the Sun to maintain a surface temperature that allows for liquid water and is massive enough to hold on to its atmosphere. The molten rock in the mantle layer above its core keeps the crust of Earth in motion (a process called plate tectonics), and the rotation of this charged material has generated a magnetic field that protects Earth from the brunt of the solar wind.

These conditions have created an environment in which life has gotten a foothold and flourished. And life has acquired enough diversity that the occasional setback (like the asteroid that may have struck Earth some 65 million years ago) might change the course of evolution of life on the planet but has not wiped it out—yet.

Mars: "That Looks Like New Mexico!"

Those of us who were glued to our television sets in 1997 when NASA shared images of the Martian surface produced by the *Mars Pathfinder* probe were struck by the resemblance of the landscape to Earth. Even the vivid red coloring of the rocky soil seemed familiar to anyone who has been to parts of Australia or even the state of Georgia—though the general landscape, apart from its color, more closely resembles desert in New Mexico. In contrast to Mercury and Venus, which are barely inclined on their axes (in fact, their axes are almost perpendicular to their orbital planes), Mars is inclined at an angle of 25.2 degrees—quite close to Earth's inclination of 23.5 degrees.

And that's only one similarity. Although Mercury and Venus move in ways very different from Earth, Mars moves through space in a way that should seem quite familiar to us. It rotates on its axis once every 24.6 hours—a little more than an Earth day—and because it is inclined much as Earth is, it also experiences familiar seasonal cycles.

The strangeness of Mercury and Venus make Mars look more similar to Earth than it really is. Generations have looked to the red planet as a kind of solar system sibling, partly believing, partly wishing, and partly fearing that life might be found there. But the fact is that life as it exists on Earth cannot exist on the other terrestrial planets.

Martian Weather Report: Cold and Thin Skies

While the atmosphere of Venus is very thick, that of Mars is very thin. It consists of about 95 percent carbon dioxide, 3 percent nitrogen, 2 percent argon, and trace amounts of oxygen, carbon monoxide, and water vapor. Although our imaginations might tend to paint Mars as a hot desert planet, it is actually a very cold, very dry place—some 50 K (on average) colder than Earth.

This section of a geometrically improved, color-enhanced version of a 360-degree pan was taken by the Imager for Mars Pathfinder *during three Martian days in 1997. Note the* Sojourner *rover vehicle and its tracks.*

(Image from NASA)

The Martian Chronicles

Percival Lowell, the son of one of New England's wealthiest and most distinguished families, was born in Boston in 1855. In the 1890s, he read a translation of an 1877 book by Giovanni Schiaparelli, the same Italian astronomer who had concluded (incorrectly, as it turned out) that Mercury's rotation was synchronized with its orbit. Reporting his observations of the surface of Mars, Schiaparelli mentioned having discovered *canali*. The word, which means nothing more than "channels" in Italian, was translated as "canals" in the translation Lowell read, and the budding astronomer, already charmed by exotic places, set off in quest of the *most* exotic of all: Mars—and whatever race of beings had excavated *canals* upon it.

Lowell dedicated his considerable family fortune to the study of the planet. He built a private observatory in Flagstaff, Arizona, and, after years of observation, published *Mars and Its Canals* in 1906. Noting that the canal network underwent seasonal changes, growing darker in the summer, Lowell theorized that technologically sophisticated beings had created the canals to transport crop-irrigation water from the Martian polar ice caps. In 1924, astronomers searched for radio signals from the planet (using a technique that anticipated the current SETI search for radio signals from the universe), but to no avail. Yet the idea of intelligent life on Mars was so ingrained in the public imagination that, on October 30, 1938, Orson Welles's celebrated radio adaptation of H. G. Wells's 1898 science fiction novel about an invasion from Mars, *War of the Worlds*, triggered national panic.

A variety of space probes have now yielded very high-resolution images of Mars, revealing the apparent canals as simply natural features, like craters or canyons. Although it is true that Mars undergoes seasonal changes, the ice caps consist of a combination of frozen carbon dioxide and water.

Why Mars Is Red

If we feel any disappointment at the loss of the Martian canals, at least we can still enjoy the image of the "angry red planet." Yet the source of the reddish hue is not the bloody spirit of the Roman god of war, but simple iron. The Martian surface contains large amounts of iron oxide, red and rusting. As *Viking 1* and *Mars Pathfinder* images revealed, even the Martian sky takes on a rust-pink tinge during seasonal dust storms.

Winds kick up in the Martian summer and blow the dust about as they play a prominent role, forming vast dunes and streaking craters. An especially large dune is found surrounding the north polar cap.

Volcanoes, Craters, and a "Grand Canyon"

The *Mariner* series of planetary probes launched in the 1960s and 1970s revealed a startling difference between the southern and northern hemispheres of Mars. The southern hemisphere is far more cratered than the northern, which is covered with wind-blown material as well as volcanic lava. Some scientists have even speculated that the smooth northern hemisphere hides a large frozen ocean.

Volcanoes and lava plains from ancient volcanic activity abound on Mars. Because the planet's surface gravity is low (0.38 that of Earth), volcanoes can rise to spectacular heights. Like Venus, Mars lacks a strong magnetic field, but, in contrast to Venus, it rotates rapidly; therefore, astronomers conclude that the core of Mars is nonmetallic, nonliquid, or both. Astronomers believe that the core of Mars has cooled and is likely solid, consisting largely of iron sulfide.

Unlike Earth, Mars failed to develop much tectonic activity (instability of the crust), probably because its smaller size meant that the outer layers of the planet cooled rapidly. Volcanic activity was probably quite intense some 2 billion years ago.

Also impressive are Martian canyons, including Valles Marineris, the "Mariner Valley," which runs some 2,500 miles (4,025 km) along the Martian equator and is as much as 75 miles (120 km) wide and, in some places, more than 4 miles (6.5 km) deep. The Valles Marineris is not a canyon in the earthly sense, because it was not cut by flowing water, but is a geological fault feature.

Astro Byte

Olympus Mons, found on Mars, is the largest known volcano in the solar system. It is 340 miles (544 km) in diameter and almost 17 miles (27 km) high.

Water, Water *Anywhere?*

Clearly visible on images produced by Martian probes are runoff and outflow channels, which are believed to be dry riverbeds, evidence that water once flowed as a liquid on Mars. Geological evidence dates the Martian highlands to 4 billion years ago, the time in which water was apparently sufficiently plentiful to cause widespread flooding. Recent theories suggest that, at the time, Mars had a thicker atmosphere that allowed water to exist in a liquid state, even at its low surface temperatures.

The Mars Global Surveyor (MGS) mission has found further geological evidence of the presence of liquid and subsurface water—evidence that has kept alive hopes that microbial life might have existed, or might even yet exist, on Mars.

Most recently, the MGS cameras have revealed two locations where water appears to have flowed briefly even in the past few years. Both locations are on the interior slopes of craters and were discovered by comparing images from the late 1990s to more recent images. For the latest images and news, check out the official MGS web site at http://mars.jpl.nasa.gov/mgs/.

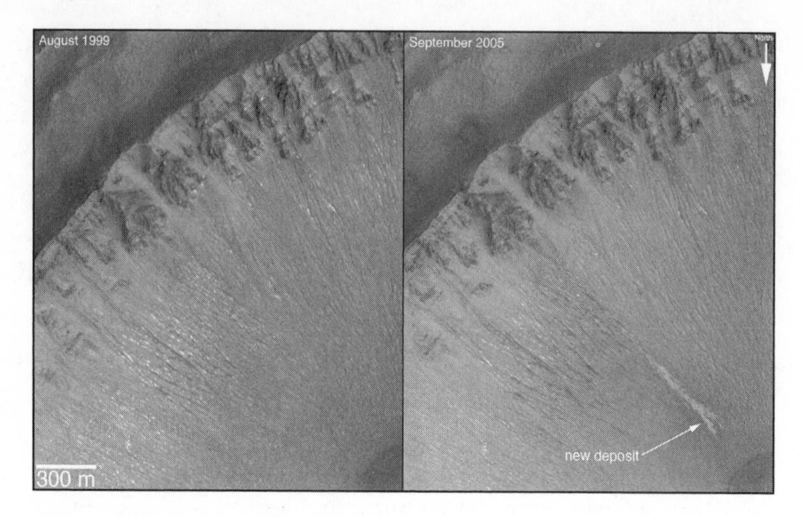

Two craters on Mars have new deposits that have formed in gully settings during the course of the Mars Global Surveyor (MGS) mission. This crater in the Centauri Montes region was first imaged by the Mars Orbiter Camera on Aug. 30, 1999. The deposit was not present at that time, so it must have formed between that date and Feb. 21, 2004.

(Image from JPL/NASA)

All Bets Are Off

Bookies no longer take bets on whether life ever existed on Mars. The odds were 1,000:1 in the 1970s and 16:1 by the start of 2004. And the results of the latest NASA mission to Mars, reported in March 2004, clearly indicated the presence of water at some time in the history of Mars, which not only removed this bet from the books but also transformed our view of the red planet as well as the solar system. The presence of water suggests the very real possibility of life.

Astro Byte

NASA launched two remote Mars Exploration Rovers (MERs), *Spirit* and *Opportunity*, on June 10 and July 7, 2003. Their primary goal was to search the Martian surface for evidence of water. The rovers landed on Mars on January 3 and January 24, 2004, and immediately got to work. They have been operational ever since, and both are now over 1,000 Martian days "past warranty."

One of the primary scientific goals of the two Mars Exploration Rovers (MERs), *Spirit* and *Opportunity*, was to determine whether or not Mars had been a wetter place in the distant past. Astronomers hoped also to determine if it had been wet for extended periods of time, long enough so life might have arisen there.

In March 2004, NASA released results from the *Opportunity* rover that landed in Merdiani Planum, a region known to be rich in hematite, a form of iron that typically forms in watery environments. The rover confirmed the presence of hematite, but that was just a start.

When the cameras sent the first images back from the landing site, JPL scientists were amazed to see that the craft had rolled to a stop in a tiny crater—a "hole in one," as some of them described it. No ordinary crater, this one contained an outcropping of rock that, to scientists, looked like something no one had ever seen before. It was Martian bedrock.

Gone were the red stone-scattered fields of the *Viking* and *Pathfinder* missions. This was an alien landscape within an alien landscape. The surface around *Opportunity* was dark and smooth, with a clear outcropping of rock at the lip of the crater. After surveying its immediate surroundings, *Opportunity* began to examine the bedrock in detail. In it, the rover's Moessbauer spectrometer detected a hydrated (water-bearing) iron sulfate mineral called jarosite. On Earth, jarosite can form in hot, acidic environments, much like hot springs.

However, scientists require multiple lines of evidence when making a dramatic claim, and the long-term presence of water on Mars is certainly dramatic. The other evidence for the presence of water includes indentations in the rock, spherules, and crossbedding. The indentations (called "vugs") are probably due to the formation of crystals in the rock when it sat for long periods in a salt-water environment. The spherules (dubbed "blueberries") have been detected at many layers in the rock, indicating that they might have formed when minerals seeped out of porous rock. Finally, tiny ridge patterns in the rock indicate either water or wind erosion over long periods of time. The small scale of the crossbedding in the outcrop rocks has geologists thinking that water is the more likely cause.

If Mars was watery for a long period of time, as these results strongly indicate, the odds for life having existed at least in one other place in the universe—indeed, in our very own solar system—go way up, and some images from the Mars Global Surveyor (MGS) mission have shown evidence of water flows recent enough to have occurred during the lifetime of the mission. For the latest information about the Mars Exploration rovers, go to www.marsrover.nasa.gov.

Martian Moons

Mars and Earth are the only terrestrial planets with moons. As we have said, our Moon is remarkably large, comparable in size to some of the moons of Jupiter. The moons of Mars, colorfully named Phobos (Fear) and Deimos (Panic), after the horses that drew the chariot of the Roman war god, were not discovered until 1877.

They are rather unimpressive as moons go, resembling large asteroids. They are small and irregularly shaped (Phobos is 17.4 miles long × 12.4 miles [28 km × 20 km] wide, and Deimos is 10 miles × 6.2 miles [16 km × 10 km]). They are almost certainly asteroids that were gravitationally captured by the planet and fell into orbit around it.

Where to Next?

Mercury, Venus, and Mars are our neighbors, the planets we know most about and perhaps feel the closest connection to. At that, however, they are still strange and inhospitable worlds. Strange? Inhospitable? Well, as Al Jolson was famous for saying, "You ain't seen nothin' yet."

The Least You Need to Know

- The terrestrial planets are Mercury, Venus, Earth, and Mars.
- Although the terrestrial planets share certain Earthlike qualities, they differ in significant ways that, among other things, make the existence of life on those planets impossible or at least highly unlikely.
- Mercury and Venus display rotational peculiarities. In the case of Mercury, we can explain its rotation by its proximity to the Sun; but the slow retrograde rotation of Venus can be best explained by the occurrence of some random event (probably a collision) early in the formation of the solar system.
- Of the terrestrial planets, only Earth has an atmosphere and environment conducive to life.
- Although Mars might look and seem familiar, its thin, icy atmosphere whipped by dust storms is a harsh environment.
- Several recent missions to Mars have provided compelling evidence of water.

Chapter 6

Bloated and Gassy: The Outer Planets

In This Chapter

- The jovian planetary line-up
- The gassy jovians
- The discovery of Uranus and Neptune
- Rotational quirks
- Atmosphere and weather
- Interior structure of the outer giants

Imagine Galileo's surprise when he pointed a telescope at the planet Jupiter in 1610. Not only was it more than a featureless point of light—it had a surface—but it was also orbited by four smaller bodies. His discovery would cause a good deal of upheaval in the way humans viewed themselves in the universe, and Galileo himself would end up in trouble with the Catholic Church. The planets are found near an imaginary arc across the sky that we call the *ecliptic*. Long before astronomers knew that the terrestrial planets shared common features, they knew two of the "wanderers" they watched were different. Although Mercury and Venus never strayed far from the Sun, and Mars moved relatively rapidly across the sky, Jupiter and Saturn moved ponderously, majestically through the stellar ocean. Long before the invention of the telescope, that motion was a clue that the outer planets—those farthest from the Sun—were unique. Mercury, Venus, and Mars might seem inhospitable, forbidding, and downright deadly, but our sister terrestrial planets have more in common with Earth than with these giants of the solar system's farthest reaches. The jovian planets are truly *other-worldly*, many times larger and more massive than Earth, yet less dense. They are balls of gas that coalesced around dense cores, accompanied by multiple moons and even rings. In this chapter, we explore these distant worlds.

The Jovian Line-Up

Jupiter was the supreme god in Roman mythology. In Middle English his name was transmuted into Jove, from which the word *jovial* comes, because, during the astrologically obsessed Middle Ages, the influence of Jupiter was regarded as the source of human happiness. Another adjective, *jovian*, usually spelled with a lowercase initial letter, means "like Jupiter." If the terrestrial planets have Earthlike ("terrestrial") qualities, all the jovian planets have much in common with Jupiter.

In addition to Jupiter, by far the largest planet in the solar system (about 300 times more massive than Earth and with a radius 11 times greater), the jovians include Saturn, Uranus, and Neptune.

Planetary Stats

The most immediately striking differences between the terrestrial and jovian worlds are in size and density. Recall our rough scale: if Earth is a golf ball 0.2 miles from the Sun, then Jupiter is a basketball 1 mile away from the Sun, and the dwarf planet Pluto is a chickpea 8 miles away. At this scale, the Sun's diameter would be as big as the height of a typical ceiling (almost 10 feet). Although the jovian planets dwarf the terrestrials, they are much less dense. Let's sum up the jovians, compared to Earth, in the following table.

We have given a gravitational force and a temperature at the "surface" of the jovian planets, but as you'll see, they don't really have a surface in the sense that the terrestrial planets do. These numbers are the values for the outer radius of their swirling atmosphere. One surprise might be that the surface gravity of Saturn, Uranus, and Neptune is very close to what we have at the surface of Earth. Gravitational force depends on two factors: mass and radius. Although these outer planets are much more massive, their radii are so large that the force of gravity at their "surfaces" is close to that of the smaller, less massive Earth.

Planet	Mass in Kilograms	Radius in Miles (and km)	"Surface" Magnetic Field (Relative to Earth)	"Surface" Gravity (Relative to Earth)	Rotation Period in Solar Days	Atmospheric Temperature in K
Earth	6.0×10^{24}	3,968 (6,380 km)	1	1	1	290
Jupiter	1.9×10^{27}	44,020 (71,400 km)	14	2.5	0.41	124
Saturn	5.7×10^{26}	37,200 (60,200 km)	0.7	1.1	0.43	97
Uranus	8.8×10^{25}	16,120 (25,600 km)	0.7	0.9	–0.72*	58
Neptune	1.0×10^{26}	15,500 (24,800 km)	0.4	1.2	0.67	59

**The rotation period of Uranus is negative because it is retrograde; like Venus, it rotates on its axis in the opposite direction from the other planets.*

Of the jovians, Jupiter and Saturn have the most in common with one another. Both are huge, their bulk consisting mainly of hydrogen and helium. During the early phases of solar system development, the outer solar system (farther from the Sun) contained more water and organic materials than the inner part, and the large mass and cooler temperatures of the outer planets meant that they were able gravitationally to hold on to the hydrogen and helium in their atmospheres.

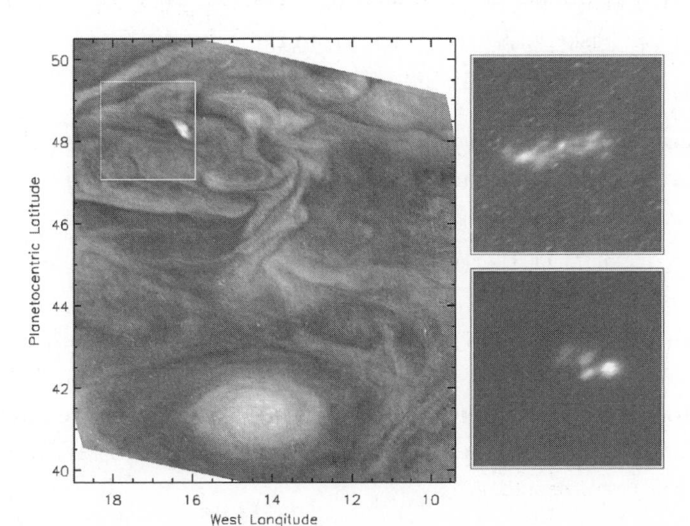

Jupiter's atmosphere is host to storms (left panel) and lightning (right panels) much like those found on Earth. These images from the Galileo *mission to Jupiter show a number of lightning strikes visible in the two images on the right, taken minutes apart.*

(Image from JPL/NASA)

Astronomer's Notebook

Why are the jovian planets so large? These planets formed from nebular material (the stuff that formed the solar system) far from the Sun, in regions that were relatively cool. The lower temperatures enabled water, ice, and other molecules to condense in the outer solar system sooner, when those same molecules were still gaseous in the region closer to the still-forming protosun. As a result, a larger quarry of solid material was available in the outer solar system to start building up into planetesimals. In effect, the outer solar system got a head start. These larger planetesimals (and their stronger gravitational fields) held on to a larger mass of hydrogen and helium, and thus the gas giants were born.

The terrestrials consist mostly of rocky and metallic materials, and the jovians primarily of lighter elements. We determine the density of a planet by dividing its mass by its volume. Although the outer planets are clearly much more massive (which, you might think, would make them more dense), they are much larger in radius, and so encompass a far greater volume. For that reason, the outer planets have (on average) a lower density than the inner planets as you can see in the following table:

Planet	Density (kg/m^3)
Earth	5,500
Jupiter	1,330
Saturn	710
Uranus	1,240
Neptune	1,670

Now, let's move on to Uranus and Neptune—distant, faint, and unknown to ancient astronomers.

Although they are both much larger than Earth, they are less than half the diameter of Jupiter and Saturn; in our scale model, they would each be about the size of a cantaloupe. Uranus and Neptune, though less massive than Saturn, are significantly more dense. Neptune is denser than Jupiter as well, and Uranus approaches Jupiter in density.

Ponder Neptune. Remember, density is equal to the mass of an object divided by its volume. Although the mass of Neptune is about 19 times smaller than that of Jupiter, its volume is 24 times smaller. Thus, we expect its density to be about $^{24}/_{19}$ or 1.3 times greater. Although we cannot yet peer beneath the atmospheric surfaces of Uranus and Neptune, the higher densities of these two planets provide a valuable clue to what's inside.

Reflecting their genesis, all the jovian planets have thick atmospheres of hydrogen and helium covering a super dense core slightly larger than Earth or Venus. The rocky cores of all four of these planets are believed to have similar radii, on the order of 4,300 to 6,200 miles (7,000 to 10,000 km); but this core represents a much smaller fraction of the full radius of Jupiter and Saturn than do the cores of smaller Uranus and Neptune—thus the higher average density of the latter two planets.

The atmospheres of the jovian planets are ancient and have probably changed little since early in the lifetime of the solar system. With their strong gravitational fields and great mass, these planets have held on to their primordial atmospheric hydrogen and helium, whereas most of these elements long ago escaped from the less massive terrestrial planets, which have much weaker gravitational pull.

But here's where things get strange. On Earth, we have the sky (and atmosphere) above and the solid ground below. In the case of the jovian planets, the gaseous atmosphere never really ends. It just becomes denser with depth, as layer upon layer presses down.

There is no "normal" solid surface to these planets! As the gases become more dense, they become liquid, which is presumably what lies at the core of the jovian worlds. When astronomers speak of the "rocky" cores of these planets, they are talking about chemical composition rather than physical state. Even on Earth, rock can be heated and pressed sufficiently to liquefy it (think of volcanic lava). It is like this deep inside the jovian atmospheres: they become increasingly dense, but never solid, surrounding a liquid core. In the case of Jupiter and Saturn, the pressures are so great that even the element hydrogen takes on a liquid metallic form.

Latecomers: Uranus and Neptune

Since ancient times, the inventory of the solar system was clear and seemingly complete: the Sun and, in addition to Earth, five planets, Mercury, Venus, Mars, Jupiter, and Saturn. Then, on March 13, 1781, the great British astronomer William Herschel, tirelessly mapping the skies with his sister Caroline, took note of what he believed to be a comet in the region of a star called H Geminorum. In fact, with the aid of a telescope, Herschel had discovered Uranus, the first new planet since ancient times.

After Uranus had been found, a number of astronomers began plotting its orbit. But something was wrong. Repeatedly, over the next half century, the planet's observed positions didn't totally coincide with its mathematically predicted positions. By the early nineteenth century, a number of astronomers began speculating that the new planet's apparent violation of Newton's laws of motion had to be caused by the influence of some undiscovered celestial body—that is, yet another planet. For the first time, Isaac Newton's work was used to identify the irregularity in a planet's orbit and to predict where another planet should be. All good scientific theories are able to make testable predictions, and here was a golden opportunity for Newton's theory of gravity.

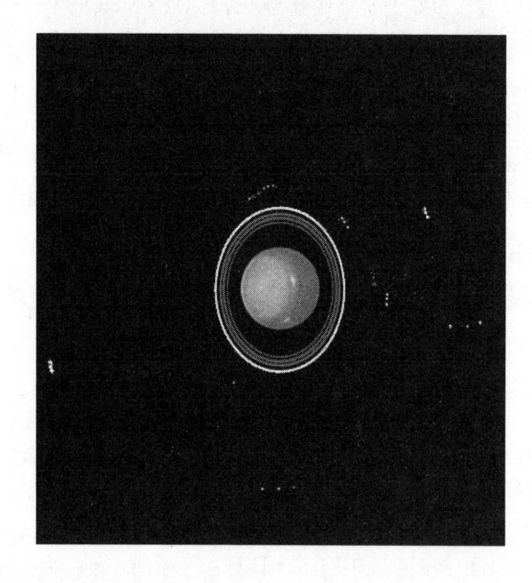

NASA's Hubble Space Telescope made this image of Uranus, including its rings, the inner moons, and distinct clouds in Uranus's southern hemisphere.

(Image from K. Seidelmann/USNO/NASA)

On July 3, 1841, John Couch Adams (1819–1892), a Cambridge University student, wrote in his diary: "Formed a design in the beginning of this week of investigating, as soon as possible after taking my degree, the irregularities in the motion of Uranus … in order to find out whether they may be attributed to the action of an undiscovered planet beyond it …." True to his word, in 1845, he sent to James Challis, director of the Cambridge Observatory, his calculations on where the new planet, as yet undiscovered, could be found. Challis passed the information to another astronomer, George Airy, who didn't get around to doing anything with the figures for a year. By that time, working with calculations supplied by another astronomer (a Frenchman named Jean Joseph Leverrier), Johann Galle, of the Berlin Observatory, found the planet that would be called Neptune. The date was September 23, 1846.

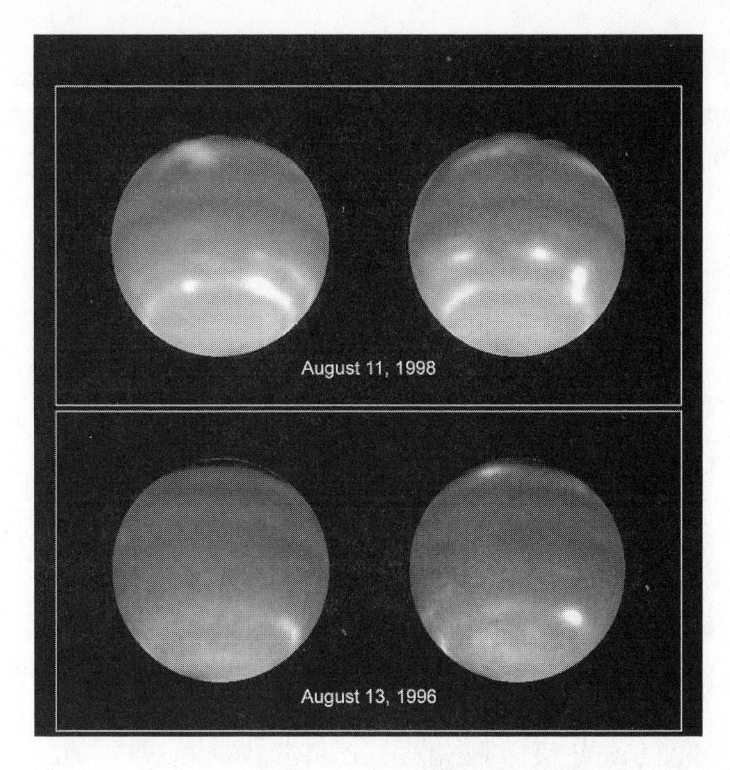

Stormy weather on Neptune. Made with a combination of observations from the Hubble Space Telescope (HST) and the NASA Infrared Telescope Facility (ITF) on Mauna Kea, Hawaii, this image shows a surface mottled with storms and torn by winds of up to 900 miles per hour.

(Image from L. Sromovsky/ UW-Madison/NASA)

Close Encounter

The dark brownish stripes across Jupiter are called *belts,* the brighter stripes, *zones.* Belts are dark, cooler regions, settling lower into the atmosphere as part of a convective cycle. Zones are regions of rising hot atmospheric gas. The bands are the result of regions of the atmosphere moving from high pressure to low pressure regions (much as they do on Earth). The rapid rotation of Jupiter confines this movement to narrow belts. The planet does have an atmospheric geography that can be mapped:

- The light-colored central band is the equatorial zone. It might appear white, orange, or yellow.
- The Great Red Spot, a hurricane that has been observed south of the equatorial zone since the invention of telescopes, has a diameter approximately twice that of Earth.
- On either side of the equatorial zone are dark bands called the north and south equatorial belts. At times, you might witness a south equatorial belt disturbance: an atmospheric storm.
- North and south of the equatorial belts are the *north* and *south temperate belts.*
- At the extreme northern and southern ends of Jupiter are the *polar regions,* which are sometimes barely perceptible and sometimes quite apparent dark areas.

Views from the *Voyagers* and *Galileo*

During the 1970s and 1980s, two *Voyager* space probes gave us unprecedented images of the jovian planets. *Voyager 1* visited Jupiter and Saturn, and *Voyager 2* added Uranus and Neptune to the list.

The *Voyager* missions also revealed volcanic activity on Io, one of Jupiter's moons. As for Saturn, a new, previously unknown system of rings emerged: several thousand ringlets. Ten additional moons were discovered orbiting Uranus, which also revealed the presence of a stronger magnetic field than had been predicted. And the Neptune flyby led to the discovery of three planetary rings as well as six previously unknown moons. The hitherto featureless blue face of the planet was resolved into atmospheric bands, as well as giant cloud streaks. As a result of the *Voyager 2* flyby, the magnetospheres of Neptune and Uranus were detected. As with the *Van Allen belts* around Earth, the magnetospheres of these planets trap charged particles (protons and electrons) from the solar wind.

def•i•ni•tion

The **Van Allen belts,** named for their discoverer, American physicist James A. Van Allen, are vast doughnut-shaped zones of highly energetic charged particles that are trapped in the magnetic field of Earth. The zones were discovered in 1958.

If only its namesake could have lived to see it! Launched in 1989, *Galileo* reached Jupiter in 1995 and began a complex 23-month orbital tour of the planet and its moons almost 400 years after the Italian astronomer first gazed on them. Among the most extraordinary of *Galileo*'s discoveries is a new ring of dust that has a retrograde (backward) orbit around Jupiter. About 700,000 miles (1,120,000 km) in diameter, this doughnut-shaped ring moves in the opposite direction of the rotating planet and its moons. The *Galileo* mission ended dramatically on September 21, 2003, when the spacecraft entered Jupiter's atmosphere. It transmitted briefly before succumbing to high temperatures and pressures.

Close Encounter

Late in July 1994, *Galileo* was in orbit on the far side of Jupiter when more than 20 fragments of Comet Shoemaker-Levy 9 plunged into the atmosphere over a 6-day period. The fragments were the result of Jupiter's tidal forces, which pulled a previously normal comet into a chain of smaller comets. Traveling at more than 40 miles per second (60 km/s), the fragments created a series of spectacular explosions in the planet's upper atmosphere, each with a force comparable to the detonation of a billion atomic bombs. So-called "black-eye" impact sites were created, the result of changes that the impacts produced in the upper atmosphere of the planet.

The View from *Cassini*

The *Cassini* spacecraft, which started its tour of Saturn and its moons in July 2004, is now well into its exploration of the planet, its moons, and its magnetosphere. In its planned four-year tour, it will complete 74 orbits of Saturn, 44 close flybys of Titan, and 8 close flybys of other satellites, including Enceladus, Phoebe, Hyperion, Dione, Rhea, and Iapetus.

The *Cassini-Huygens* mission has changed our understanding of the ringed planet and confirmed many hypotheses about the nature of its mysterious moon, Titan. In the spring of 2007, radar imagers on NASA's *Cassini* spacecraft found evidence of seas on Titan, most likely filled with liquid methane or ethane. Several of these seas are much larger than the Great Lakes of North America.

This image of Saturn and its rings is from the Cassini *wide-angle camera, taken from a distance of approximately 999,000 kilometers (621,000 miles) on May 4, 2005. The spacecraft was only a few degrees above the ring plane.*

Rotation: A New Twist

With all the bands and surface features of the biggest jovian planets, you'd think it would be relatively easy to calculate rotation rates "by eye." One could just look for a prominent surface feature and time how long it takes that feature to make one trip around.

Well, it's not that easy. Because these planets lack solid surfaces, different features on the surface actually rotate at differing rates. This differential rotation is not dramatic in the case of Jupiter, whose equatorial region rotates only slightly faster than regions at higher latitudes. East-west winds move at about 190 miles per hour (300 km/h) in Jupiter's equatorial regions and at a zippy 800 miles per hour (1,300 km/h) in the equatorial regions of Saturn. It turns out that the best way to clock the rotation rates of these planets is not to look at their atmospheres but to measure something tied to the planets' cores. The periods of fluctuation in the radio emission (which arise from the core-generated magnetic fields) are taken to be the "true" rotation rate.

Although Neptune and Saturn are slightly tipped on their axes similar to Earth (30, 27, and 23.5 degrees, respectively), Jupiter's axis is nearly perpendicular to the plane of its orbit; the planet tilts from the perpendicular a mere 3 degrees.

The true oddball in this respect is Uranus, which tilts 98 degrees, in effect lying on its side. The result of this peculiarity is that Uranus has the most extreme seasons in the solar system. While one pole experiences continuous daylight for 42 Earth years at a stretch, the other is plunged into an equal period of darkness.

Stormy Weather

Jupiter's spectacular surface features belie a turbulent atmosphere. In addition to a prevailing eastward and westward wind flow called *zonal flow*, many smaller-scale weather patterns exist as evidenced by such features as the Great Red Spot.

The Great Red Spot

The British scientist Robert Hooke (1635–1703) first reported the Great Red Spot, which is a storm, a swirling hurricane or whirlpool, of gigantic dimensions (twice the size of Earth), at least 300 years old. It rotates once every six days and is accompanied by other smaller storms. Neptune has a similar storm called the Great Dark Spot.

How could a storm last for three centuries or more? We know from our experience on Earth that hurricanes form over the ocean and might remain active there for days or weeks. When they move over land, however, they are soon spent (albeit often destructively); the land mass disrupts the flow pattern and removes the source of energy. On Jupiter, however, there is no land. Once a storm starts, it continues indefinitely until a larger storm disrupts it. The Great Red Spot is currently the reigning storm on the planet.

Bands of Atmosphere

The atmospheric bands that are Jupiter's most striking feature are the result of *convective motion* and zonal wind patterns. Warm gases rise, while cooler gases sink. The location of particular bands appears to be associated with the wind speed on Jupiter at various latitudes.

Anyone who watches an earthly television weather forecast is familiar with high-pressure and low-pressure areas. Air masses move from high-pressure regions to low-pressure regions. But we never see these regions on Earth as regular zones or bands that circle the planet. That's because Earth doesn't rotate nearly as fast as Jupiter. The rapid rotation of the gas giant spreads the regions of high and low pressure out over the entire planet.

def•i•ni•tion

Convective motion is any flow pattern created by the rising movement of warm gases (or liquids) and the sinking movement of cooler gases (or liquids). Convective motion is known to occur in planetary atmospheres, the Sun's photosphere, and the tea kettle boiling water on your stove.

Layers of Gas

On July 13, 1995, near the start of its mission, *Galileo* released an atmospheric probe, which plunged into Jupiter's atmosphere and transmitted data for almost an hour before intense atmospheric heat and pressure destroyed it. After analysis of this data (and earlier data from *Voyager*), astronomers concluded that Jupiter's atmosphere is arranged in distinct layers. Because there is no solid surface to call sea level, the troposphere (the region containing the clouds we see) is considered zero altitude, and the atmosphere is mapped in positive and negative distances from this.

Just above the troposphere is a haze layer, and just below it are white clouds of ammonia ice. Temperatures in this region are 125–150 K. Starting at about 40 miles

Astro Byte

The core temperature of Jupiter must be very high, perhaps 40,000 K. Astronomers speculate that the core diameter is about 12,500 miles (20,000 km), about the size of Earth. As the jovian planets collapsed, part of their gravitational energy was released as heat. Some of this heat continues to be released, meaning that Jupiter, Saturn, and Neptune have internal heat sources.

(60 km) below the ammonia ice level is a cloud layer of ammonium hydrosulfide ice, in which temperatures climb to 200 K. Below this level are clouds of water ice and water vapor, down to about 60 miles (100 km). Farther down are the substances that make up the interior of the planet: hydrogen, helium, methane, ammonia, and water, with temperatures steadily rising the deeper we go. Molecules at these depths can be probed with observations at radio wavelengths.

Saturnine Atmosphere

Saturn's atmosphere is similar to Jupiter's—mostly hydrogen (92.4 percent) and helium (7.4 percent) with traces of methane and ammonia; however, its weaker gravity results in thicker cloud layers that give the planet a more uniform appearance, with much subtler banding, than Jupiter. Temperature rises much more slowly as a function of depth in the atmosphere on Saturn.

The Atmospheres of Uranus and Neptune

Unmanned space vehicles have not probed the atmospheres of Uranus and Neptune, but these atmospheres have been studied spectroscopically from Earth, revealing that, like Jupiter and Saturn, they are mostly hydrogen (about 84 percent) and helium (about 14 percent). Methane makes up about 3 percent of Neptune's atmosphere and 2 percent of Uranus's, but ammonia is far less abundant on either planet than on Jupiter and Saturn. Because Uranus and Neptune are colder and have much lower atmospheric pressure than the larger planets, any ammonia present is frozen. The lack of ammonia in the atmosphere and the significant presence of methane give both Uranus and Neptune a bluish appearance because methane absorbs red light and reflects blue. Uranus, with slightly less methane than Neptune, is blue-green while Neptune is quite blue.

Uranus reveals almost no atmospheric features. The few present are submerged under layers of haze. Neptune, as seen by *Voyager 2*, reveals more atmospheric features and even some storm systems, including a Great Dark Spot, an area of storm comparable in size to Earth. Discovered by *Voyager 2* in 1989, the Great Dark Spot had vanished by the time the Hubble Space Telescope observed the planet in 1994.

Inside the Jovians

How do you gather information about the interior of planets that lack a solid surface and are so different from Earth? You combine the best observational data you have with testable, constrained speculation known as theoretical modeling. Doing just this, astronomers have concluded that the interiors of all four jovians consist largely of the elements found in their atmospheres: hydrogen and helium. Deeper in the planets, the gases, at increasing pressure and temperature, become liquid but never solid.

In the case of Jupiter, astronomers believe that the hot liquid hydrogen is transformed from molecular hydrogen to metallic hydrogen and behaves much like a molten metal, in which electrons are not bound to a single nucleus but move freely, conducting electrical charge. As you shall see in just a moment, this state of hydrogen is likely related to the creation of Jupiter's powerful magnetosphere—the result of its strong magnetic field. Saturn's internal composition is doubtless similar to Jupiter's although its layer of metallic hydrogen is probably proportionately thinner, and its core is slightly larger. Temperature and pressure at the Saturnine core are certainly less extreme than on Jupiter.

Uranus and Neptune are believed to have rocky cores of similar size to those of Jupiter and Saturn surrounded by a slushy layer consisting of water clouds and, perhaps, the ammonia that is largely absent from the outer atmosphere of these planets. Because Uranus and Neptune have significant magnetospheres, some scientists speculate that the ammonia might create an electrically conductive layer, needed to generate the detected magnetic field. Above the slushy layer is molecular hydrogen. Without the enormous internal pressures present in Jupiter and Saturn, the hydrogen does not assume a metallic form.

The Jovian Magnetospheres

Jupiter's magnetosphere is the most powerful in the solar system. Its extent reaches some 18,600,000 miles (30 million km) north to south. Saturn has a magnetosphere that extends about 600,000 miles (1 million km) toward the Sun. The magnetospheres of Uranus and Neptune are smaller, weaker, and (strangely) offset from the gravitational center of the planets.

The rapid rate of rotation and the theorized presence of electrically conductive metallic hydrogen inside Jupiter and Saturn account for the strong magnetic fields of these planets. Although Uranus and Neptune also rotate rapidly, it is less clear what internal material generates the magnetic fields surrounding these planets because they are not thought to have metallic hydrogen in their cores. With charged particles trapped by their magnetospheres, the jovian planets experience *Aurora Borealis*, or "Northern Lights," just as we do here on Earth. These "lights" occur when charged particles escape the magnetosphere and spiral along the field lines onto the planet's poles. The Hubble Space Telescope has imaged such auroras at the poles of Jupiter and Saturn.

The Least You Need to Know

- The jovian planets are Jupiter, Saturn, Uranus, and Neptune.
- Although they are the largest, most massive planets in the solar system, these planets are on average less dense than the terrestrial planets, and their outer layers of hydrogen and helium gas cover a dense core.
- Neptune was discovered in 1846 because astronomers were searching for an explanation of Uranus's slightly irregular orbit. Newton's theory of gravity provided an explanation—the mass of another planet.
- The jovian planets all have in common thick atmospheres, ring systems, and strong magnetic fields.
- The missions to the outer planets, *Voyager 1*, *Voyager 2*, and *Galileo*, have had a huge impact on increasing our understanding of the jovian planets with the *Cassini-Huygens* mission greatly expanding our understanding of Saturn and its moons.

Chapter 7

The Moon, Moons, and Rings

In This Chapter

- Observing the Moon
- Lunar geology
- The jovian rings
- The large moons of Jupiter and the other jovians
- Distant dwarves: Pluto and Charon

Our own Moon is still the only celestial body other than Earth where human beings have stood, first doing so nearly 30 years ago when Neil Armstrong stepped onto the lunar surface. What have we learned from the information we brought back from the Apollo missions? Why does the Moon fascinate us? In this chapter, we explore our Moon and the moons and rings of the outer planets.

What If We Had No Moon?

It seems like a reasonable question to ask. What if we had no moon? Would it matter? What has the Moon done for us lately?

It turns out that the presence of a large moon such as we have is unusual for a terrestrial planet. Mercury and Venus have no moons at all, and Mars has two tiny ones, Phobos and Deimos. To have a moon roughly ⅓ the size of the planet it orbits is unique in the inner solar system. Our Moon, for example, is as large as some of the moons of the giant gas planets in the outer solar system. If there were no Moon, we would have no ocean tides, and the rotation rate of Earth would not have slowed to its current 24 hours. It is thought that early in its life Earth rotated once every six hours.

The Moon also appears to stabilize the rotational axis of Earth. By periodically blocking the light from the Sun's photosphere, the Moon gives us a view of the outer layers of the Sun's atmosphere, and it also gave early astronomers clues to the relative locations of objects in the solar system.

Close Encounter

Did you ever wonder why the Sun and the Moon are the same size in the sky? In actual physical size, the Sun dwarfs the Moon, but the Sun is so much farther away that the two appear the same size. Try this: hold a dime and a quarter right in front of your face. Now hold them farther away from your face, adjusting the distance of either until both appear to be the same size. Which do you have to hold farther away for this to happen? The bigger one, of course! We happen to be on Earth at a time when the Moon exactly blocks the light from the Sun's photosphere during solar eclipses. As the Moon slowly drifts away from Earth, it will get smaller and smaller in the sky, and people will be entitled eventually to grumble, "Solar eclipses just aren't what they used to be."

What Galileo Saw

What Galileo saw when he trained a telescope on the Moon's surface, noting that it was rough and mountainous, conflicted with existing theories that the surface was glassy smooth. He closely studied the terminator (the boundary separating day and night) and noted the shining tops of mountains. Using simple geometry, he calculated the height of some of them based on the angle of the Sun and the estimated length of shadows cast. Galileo overestimated the height of the lunar mountains he observed, but he did conclude rightly that their heights were comparable to earthly peaks.

Noticing mountains and craters on the Moon was important because it helped Galileo conclude that the Moon was fundamentally not all that different from Earth. It had mountains, valleys, and even what were called seas—in Latin, *maria*, although there is no indication that Galileo or anyone else maintained after telescopic observations that the maria were in fact water-filled oceans.

Contending that the Moon resembled Earth was not a small thing in 1609 because such a statement implied that there was nothing supernatural or special about the Moon or, by implication, about the planets and the stars either. Followed to its conclusion, the observation implied that there was perhaps nothing divine or extraordinary about Earth itself. Earth was one of many bodies in space, like the Moon and the other planets.

What You Can See

Even if you don't have a telescope, you can make some very interesting lunar observations. You might, for example, try tracking the Moon's daily motion against the background stars. Because the Moon travels 360 degrees around Earth in 27.3 days, it will travel through about 13 degrees in 24 hours or about half a degree (its diameter) every hour.

The telescope through which Galileo Galilei made his remarkable lunar observations was a brand-new and very rare instrument in 1609, but you can easily surpass the quality of his observations with simple binoculars.

No other celestial object is so close to us. Being this close, the Moon provides the most detailed images of an extraterrestrial geography that you will ever see through your own instrument.

Take the time to observe the Moon through all of its phases. On clear nights when the Moon is about three or four days "old," Mare Crisium and other vivid features—including the prominent craters Burckhardt and Geminus—become dramatically visible. You can also begin to see Mare Tranquilitatis, the Sea of Tranquility, on which *Apollo 11*'s lunar module, *Eagle*, touched down on July 20, 1969.

At day seven, when the Moon is at its first quarter, mountains and craters are most dramatically visible. Indeed, this is the optimum night for looking at lunar features in their most deeply shadowed relief.

As the Moon enters its waxing gibbous phase beyond first quarter, its full, bright light is cheerful, but it's so bright that it actually becomes more difficult to make out sharp details on the lunar surface, although you do get great views of the eastern maria, the lunar plains.

Past day 14, the Moon begins to wane as the sunset terminator moves slowly across the lunar landscape. At about day 22, the Apennine Mountains are clearly visible. It was these mountains that Galileo studied most intensely, attempting to judge their heights by the shadows they cast.

During the late waning phase of the Moon, moonrise comes later and later at night as the Moon gradually catches up with the Sun in the sky. By the time the Moon passes day 26, it is nothing but a thin crescent of light present in the predawn sky. The new Moon follows, and as the Moon overtakes the Sun, the crescent reappears (on the other side of the Moon at sunset), and it begins to wax again.

Cold, Hard Facts About a Cold, Hard Place

The Moon is Earth's only natural satellite and, as noted previously, is a very large satellite for a planet as small as Earth. The planet Mercury is only slightly larger than the Moon. The mean distance between Earth and the Moon, as it orbits Earth from west to east, is 239,900 miles (386,239 km). The Moon is less than one third the size of Earth, with a diameter of about 2,160 miles (3,476 km) at its equator. Moreover, it is much less massive and less dense than Earth—$\frac{1}{80}$ as massive, with a density of 3.34 g/cm^3, in contrast to 5.52 g/cm^3 for Earth. Here's one way to think about relative sizes and distances: if Earth were the size of your head, the orbiting Moon would be the size of a tennis ball 30 feet away.

Because the Moon is so much less massive than Earth and about a third as big, its surface gravity is about one sixth that of our planet. That's why the *Apollo* astronauts could skip and jump as they did, even wearing those heavy space suits. If you weigh 160 pounds on Earth's surface, you would weigh only 27 pounds on the Moon. This apparent change would give you the feeling of having great strength because your body's muscles are accustomed to lifting and carrying six times the load that burdens them on the Moon. Of course, your mass—how much matter is in you—does not change. If your mass is 60 kilograms (kg) on Earth, it will still be 60 kilograms on the Moon.

The Moon is in a *synchronous orbit* around Earth; that is, it rotates once on its axis every 27.3 days, which is the same time it takes to complete one orbit around Earth. Thus synchronized, we see only one side of the Moon.

It's a Moon!

No one can say with certainty how the Moon was formed (we weren't there!), but astronomers have advanced four major theories. Let's look at each in turn.

Astro Byte

The Moon is much smaller in the night sky than you might think. Cut a small circle (about 1⁄5 inch in diameter) from a piece of paper. Hold that paper out at arm's length. That is how large the Moon is in the night sky. And don't let anyone tell you that a harvest moon (a full moon in September) is any larger than any other full moon. The Moon might be lower in the sky in autumn (as is the Sun), so it might look larger, but its size is unchanged.

A Daughter?

The oldest of the four theories speculates that the Moon was originally part of Earth, which was somehow spun off as a rapidly rotating, partially molten, newly forming planet.

Once prevalent, this theory (sometimes referred to as the fission theory) has largely been rejected because it does not explain how the proto-Earth could have been spinning with sufficient velocity to eject the material that became the Moon. Moreover, it is highly unlikely that such an ejection would have put the Moon into a stable Earth orbit.

A Sister?

Another theory holds that the Moon formed separately near Earth from the same material that made up Earth. In effect, Earth and the Moon formed as a double-planet system.

This theory seemed quite plausible until lunar rock samples were recovered, revealing that the Moon differs from Earth not only in density but also in composition. If the two bodies had formed out of essentially the same stuff, why would their compositions be so different?

A Captive?

A third theory suggests that the Moon was formed independently and far from Earth but was later captured by Earth's gravitational pull when it came too close.

This theory can account for the differences in composition between Earth and the Moon, but it doesn't explain how Earth could have gravitationally captured such a large Moon. Indeed, attempts to model this scenario with computer simulations have failed; the rogue moon can't be captured gravitationally unless the two collide. Moreover, although the theory accounts for some of the chemical *differences* between Earth and the Moon, it does not explain the many chemical *similarities* that also exist.

A Fender Bender?

Today's favored theory combines elements of the daughter theory and the capture theory in something called the impact theory. Most astronomers now believe that a very large object, roughly the size of Mars, collided with Earth when Earth was still molten and forming. Assuming the impact was a glancing one, it is suggested that a piece of shrapnel from Earth and the remnant of the other planetesimal (a planet in an early stage of formation) were ejected and then slowly coalesced into a stable orbit that formed the Moon.

This model is also popular because it explains some unique aspects of Earth (the "tip" of its rotational axis, for instance) and the Moon. In the impact model, it is further theorized that most of the iron core of the Mars-sized object would have been left behind on Earth, eventually to become part of Earth's core, while the material that would coalesce into the Moon retained less of this metallic component. This model therefore explains why Earth and the Moon share similar mantles (outer layers) but apparently differ in core composition.

Give and Take

Isaac Newton proposed that every object with mass exerts a gravitational pull or force on every other object with mass in the universe. Well, Earth is much more massive (80 times more) than the Moon, which is why the Moon orbits us and not we it. If you want to get technical, we both actually orbit an imaginary point called the center of mass. However, the Moon *is* sufficiently massive to make the effects of its gravitational field felt on Earth.

Anyone who lives near the ocean is familiar with tides. Coastal areas experience two high and two low tides within any 24-hour period. The difference between high and low tides is variable, but, out in the open ocean, the difference is somewhat more than 3 feet. If you've ever lifted a large bucket of water, you know how heavy water is. Imagine the forces required to raise the level of an entire ocean 3 or more feet!

What force can accomplish this? The *tidal force* of gravity exerted by the Moon on Earth and its oceans.

The Moon and Earth mutually pull on each other, Earth's gravity keeping the Moon in its orbit and the Moon's gravity causing a small deformity in Earth's shape. This deformity results because the Moon does not pull equally on all parts of Earth. It exerts more force on parts of Earth that are closer and less force on parts of Earth that are farther away.

Newton told us that gravitational forces decrease with the square of the distance. These differential or tidal forces are part of the cause of Earth's slightly distorted shape—it's ovoid rather than a perfect sphere—and they also make the oceans flow to two locations on Earth: directly below the Moon and on the opposite side. This flow causes the oceans to be deeper at these two locations, which are known as the *tidal bulges.* The Moon pulls the entire Earth into a somewhat elongated football shape, but the oceans, being less rigid than the earth, undergo a greater degree of deformity.

Interestingly, the side of Earth farthest from the Moon at any given time also exhibits a tidal bulge. This is because Earth experiences a stronger gravitational pull than the ocean on top of it, and Earth is "pulled away" from the ocean on that side. As Earth rotates (once every 24 hours) beneath the slower-orbiting Moon (once every 27.3 days), the forces exerted on the water cause high and low tides to move across the face of Earth.

The tides of largest range are the spring tides, which occur at new moon, when the Moon and the Sun are in the same direction, and at full moon, when they are in opposite directions. The tides of smallest range are the neap tides, which occur when the Sun and the Moon are at 90 degrees to one another in the sky. Tides affect you every day, of course, especially if you happen to be a sailor or a fisherman. Earth's rotation is slowing down at a rate that increases the length of a day by approximately 2 milliseconds (2⁄1,000 of a second) every century. Over millions of years, this slowing effect adds up. Five hundred million years ago, a day was a little over 21 hours long, and a year (one orbit of the Sun) was packed with 410 days. When a planetesimal plowed into Earth early in the history of the solar system, Earth was rotating once every six hours. (And you think there aren't enough hours in the day now!)

Green Cheese?

On any night the Moon is visible, the large, dark maria are also clearly visible. These vast plains were created by lava spread during a period of the Moon's evolution marked by intense volcanic activity. The lighter areas visible to the naked eye are called highlands. Generally, the highlands represent the Moon's surface layer, its *crust*, while the maria consist of much denser rock representative of the Moon's lower layer, its *mantle*. The surface rock is fine-grained, as was made dramatically apparent by the image of the first human footprint on the Moon. The maria resemble terrestrial *basalt*, created by molten mantle material that, through volcanic activity, swelled through the crust.

This Place Has Absolutely No Atmosphere

The mass of the Moon is insufficient for it to have held on to its atmosphere. As the Sun heated up the molecules and atoms in whatever thin atmosphere the Moon might have once had, they drifted away into space. With no atmosphere, the Moon has no weather, no erosion—other than what asteroid impacts cause—and no life. Although astronomers once thought the Moon had absolutely no water, recent robotic lunar missions have shown there might be water (in the form of ice) in the permanent shadows of the polar craters.

An Apollo 11 *astronaut left this footprint in the lunar dust. Unless a stray meteoroid impact obliterates it, the print will last for millions of years on the waterless, windless Moon.*

(Image from NASA)

A Pocked Face

Look at the Moon through even the most modest of telescopes—as Galileo did—or binoculars, and first and foremost you are impressed by the craters that pock its surface.

Most craters are the result of asteroid and meteoroid impacts. Only about a hundred craters have been identified on Earth, but the Moon has thousands, great and small. Was the Moon just unlucky? No. Many meteoroids that approach Earth burn up in our atmosphere before they strike the ground. And the traces of those that do strike the ground are gradually covered by the effects of water and wind erosion, as well as by plate tectonics. Without an atmosphere, the Moon has been vulnerable to whatever has come its way, preserving a nearly perfect record of every impact it has ever suffered.

Meteoroid collisions release terrific amounts of energy. Upon impact, heat is generated, melting and deforming the surface rock, while pushing rock up and out and creating an *ejecta blanket* of debris, including large boulders and dust. This ejected material covers much of the lunar surface.

def•i•ni•tion

An **ejecta blanket** is the debris displaced by a meteoroid impact.

And What's Inside?

The Moon is apparently as dead geologically as it is biologically. Astronauts have left seismic instruments on the lunar surface, which have recorded only the slightest seismic activity, barely perceptible, in contrast to the exciting (and sometimes terribly destructive) seismic activity common on Earth and some other bodies in the solar system (such as Io, a moon orbiting Jupiter).

Astronomers believe, then, that the interior of the Moon is uniformly dense, poor in heavy elements (such as iron) but high in silicates. The core of the Moon, about 250 miles (402 km) in diameter, might be partially molten. Around this core is probably an inner mantle, perhaps 300 miles (483 km) thick, consisting of semisolid rock, and around this layer, a solid outer mantle some 550 miles (885 km) thick. The lunar crust is of variable thickness, ranging from 40 to 90 miles (64 to 145 km) thick.

The Moon is responsible for everything from Earth's tides to the length of our day and perhaps the presence of seasons. Think of that the next time you see the Moon shining peacefully over your head.

Lord of the Rings

Our Earth has one of the most dazzling moons in the solar system, but when it comes to spectacular rings, we must bow to Saturn.

Looking from Earth

Galileo's telescope, a wondrous device in the early seventeenth century, would be no match for even a cheap amateur instrument today. When he first observed Saturn, all Galileo could tell about the planet was that it seemed to have "ears." He speculated that this feature might be topographical, great mountain ranges of some sort. Or, he thought, perhaps Saturn was a triple planet system, with the "ears" as outrigger planets. It wasn't until a half-century after Galileo's speculations, in 1656, that Christian Huygens of the Netherlands was able to make out this feature for what it was: a thin ring encircling the planet. A few years later, in the 1670s, the Italian-born French astronomer Gian Domenico Cassini (1625–1712) discovered the dark gap between what are now called rings A and B. We call this gap the *Cassini division*.

In all, six major rings, each lying in the equatorial plane of Saturn, have been identified, of which three, in addition to the Cassini division and a subtler demarcation called the Encke division, you can see from Earth with a good telescope. With a typical amateur instrument you should be able to see (at the very least) ring A (the outermost ring), the Cassini division, and inside the Cassini division, ring B.

The rings most readily visible from Earth are vast, the outer radius of the A ring stretching more than 84,800 miles.

Big as they are, the rings are also very thin—in places only about 65 feet (20 m) thick. If you wanted to make an accurate scale model of the rings and fashioned them to the thickness of this sheet of paper, they would have to be a mile wide to maintain proper scale.

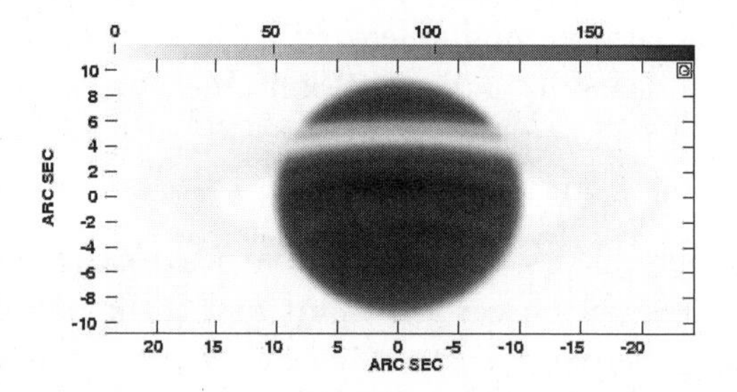

Saturn and its rings as seen with the Very Large Array at 3.6 cm. The cooler ring system is seen in absorption (light) against the dark emission from the upper atmosphere of the planet.

(Image from B. Dunn/NRAO/AUI)

Where do the rings come from? There are two ways to try to answer this question, and both involve the gravitational field of the host planet. First, the rings might be the result of a shattered moon. According to this theory, a satellite could have been orbiting too close to the planet and been torn apart by tidal forces (the same sort of forces that pulled comet Shoemaker-Levy 9 to pieces), or the moon might have been shattered by a collision. In either case, the pieces of the former moon continued to orbit the planet, but now as fragmentary material.

The other possibility is that the rings are material left over from the formation of the planet itself, material that was never able to coalesce into a moon due to the strong gravitational field of the host planet.

Up Close and Personal: *Voyager*

Zooming in close, the *Voyager* probes told us much more about the rings than we could have discovered from our earthly perspective. First, data from *Voyager* confirmed that the rings are indeed made up of particles, primarily of water ice. *Voyager* also revealed additional rings, invisible from Earth. The F ring, for example, is more than twice the size of the A ring, stretching out to 186,000 miles.

The D ring is the innermost ring—closer to the planet than the innermost ring visible from Earth, the C ring. F and E are located outside the A ring.

But these additional rings are only part of what *Voyager* told us. *Voyager 2* revealed that the six major rings are composed of many thousands of individual ringlets, which astronomers liken to ripples or waves in the rings.

Astronomer's Notebook

The biggest gap in Saturn's rings, the Cassini division, is the result not of moonlets, but of the gravitational influence of Mimas, Saturn's innermost moon, which deflected some of the ring's particles into different orbits, creating a gap large enough to be visible from Earth.

More Rings on the Far Planets

During a 1977 Earth-based observation of Uranus in the course of a stellar occultation (the passage of Uranus in front of a star), the star's light dimmed several times before disappearing behind the planet. That dimming of the star's light revealed the presence of nine thin, faint rings around the planet. *Voyager 2* revealed another pair. Uranus's rings are very narrow—most of them less than 6 miles (10 km) wide—and are kept together by the kind of shepherd satellites found outside of Saturn's F ring. Neptune has faint rings similar to those of Uranus.

On the Shoulders of Giants

One of the key differences between the terrestrial and many jovian planets is that, although the terrestrials have few if any moons, the jovians have many: 62 (at least) for Jupiter, 60 for Saturn, 27 for Uranus, and 13 for Neptune. Of these known moons, only six are classified as large bodies, comparable in size to Earth's moon. Our own moon is all the more remarkable when compared to the moons of the much larger jovian planets. It is larger than all of the known moons except for Ganymede, Titan, Callisto, and Io. The largest jovian moons (in order of decreasing radius) are …

- **Ganymede,** orbiting Jupiter; approximate radius: 1,630 miles (2,630 km).
- **Titan,** orbiting Saturn; approximate radius: 1,600 miles (2,580 km).
- **Callisto,** orbiting Jupiter; approximate radius: 1,488 miles (2,400 km).
- **Io,** orbiting Jupiter; approximate radius: 1,130 miles (1,820 km).
- **Europa,** orbiting Jupiter; approximate radius: 973 miles (1,570 km).
- **Triton,** orbiting Neptune; approximate radius: 856 miles (1,380 km).

It is interesting to compare these to Earth's moon, with a radius of about 1,079 miles (1,740 km), and the dwarf planet Pluto, smaller than them all, with a radius of 713 miles (1,150 km).

The rest of the moons are either medium-sized bodies—with radii from 124 miles (200 km) to 465 miles (750 km)—or small bodies, with radii of less than 93 miles (150 km). In recent years, we have discovered many small bodies orbiting Jupiter and Saturn. Many of the moons are entirely or mostly composed of water ice, and some of the smallest bodies are no more than irregularly shaped rock and ice chunks.

Faraway Moons

Thanks to the *Voyager*, *Galileo*, and *Cassini* space probes, we have some remarkable images and data about the moons at the far reaches of the solar system. The so-called Galilean moons of Jupiter, Saturn's Titan, and Neptune's Triton have received the most attention because they are the largest.

Jupiter's Four Galilean Moons

The four large moons of Jupiter are very large, ranging in size from Europa, only a bit smaller than Earth's moon, to Ganymede, which is larger than the planet Mercury. Certainly, they are large enough to have been discovered even through the crude telescope of Galileo Galilei, after whom they have been given their group name. In his notebooks, Galileo called the moons simply I, II, III, and IV. Fortunately, they were eventually given more poetic names: Io, Europa, Ganymede, and Callisto, drawn from Roman mythology. These four are, appropriately, the attendants serving the god Jupiter.

Io is closest to Jupiter, orbiting at an average distance of 261,640 miles (421,240 km); Europa comes next (416,020 miles or 669,792 km); then Ganymede (663,400 miles or 1,068,074 km); and finally Callisto (1,165,600 miles or 1,876,616 km). Intriguingly, data from *Galileo* suggests that the core of Io is metallic and its outer layers rocky—much like the planets closest to the Sun. Europa has a rocky core, with a covering of ice and water. The two outer large moons, Ganymede and Callisto, also have icy surfaces surrounding rocky cores.

This pattern of decreasing density with distance from the central body mimics that of the solar system at large, in which the densest planets, those with metallic cores, orbit nearest the Sun, while those composed of less dense materials orbit farthest away. This similarity is no mere coincidence, and we can use it to investigate how the Jupiter "system" formed and evolved. Let's look briefly at each of Jupiter's large moons.

Because of our own moon, we are accustomed to thinking of moons generally as geologically dead places. But nothing could be further from the truth in the case of Io, which has the distinction of being the most geologically active object in the entire solar system.

Io's spectacularly active volcanoes continually spew lava, which keeps the surface of the moon relatively smooth—any craters are quickly filled in—but also angry-looking, vivid orange and yellow, and sulfurous. In truth, Io is much too small to generate the kind of heat energy that produces volcanism (volcanic activity); however, orbiting as close as it does to Jupiter, it is subjected to the giant planet's tremendous gravitational field, which produces tidal forces that stretch the planet from its spherical shape and create its geologically unsettled conditions. Think about what happens when you rapidly squeeze a small rubber ball. The action soon makes the ball quite warm.

The forces Jupiter exerts on Io are analogous to this, but on a titanic scale. By the way, don't invest in an Io globe for your desk. Its surface features change faster than political boundaries on Earth!

In contrast to Io, Europa is a cold world—but probably not an entirely frozen one, and perhaps, therefore, not a biologically dead one. Images from *Galileo* suggest that Europa is covered by a crust of water ice, which is networked with cracks and ridges. It is possible that beneath this frozen crust is an ocean of *liquid* water (not frozen water or water vapor).

This ice is on the surface of Europa. The smooth dark regions might be areas where water has welled up from underneath the "ice shelf" that covers the moon.

(Image from NASA/JPL)

Ganymede is the largest moon in the solar system (bigger than the planet Mercury). Its surface shows evidence of subsurface ice liquefied by the impact of asteroids and then refrozen. Callisto is smaller but similar in composition. Both are ancient worlds of water ice, impacted by craters. There is little evidence of the current presence of liquid water on these moons.

Titan: Saturn's Highly Atmospheric Moon

If Io is the most geologically active moon in the solar system and Ganymede the largest, Saturn's Titan enjoys the distinction of having the most substantial atmosphere of any moon. No wispy, trace covering, Titan's atmosphere is mostly nitrogen (90 percent) and argon (nearly 10 percent) with traces of methane and other gases in an atmosphere thicker than Earth's. Earth's atmosphere consists of 78 percent nitrogen, 21 percent oxygen, and 1 percent argon. Surface pressure on Titan is about 1.5 times that of Earth. But its surface is very cold, about 90 K. Remember 90 K is –183°C!

Titan's atmosphere prevents any visible-light view of the surface, although astronomers speculate that the interior of Titan is probably a rocky core surrounded by ice, much like that of Ganymede and Callisto. Because Titan's temperature is lower than that of Jupiter's large moons, it has retained its atmosphere. The presence of an atmosphere thick with organic molecules (carbon monoxide, nitrogen compounds, and various hydrocarbons have been detected in the upper atmosphere) has led to speculation that Titan could possibly support some form of exotic life. *Cassini-Huygens* arrived at Saturn in 2004; since arrival, it has had many close flybys of Titan, and the *Huygens* lander came to rest on the surface of the planet in 2005. Most recently, astronomers have found strong evidence of the presence of ethane or methane lakes, some larger than the Great Lakes of North America.

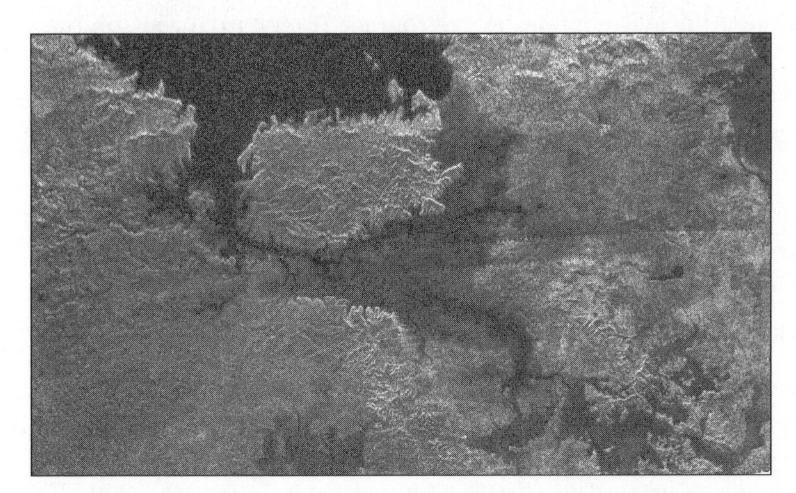

This radar image, obtained by the Cassini *probe during a near-polar flyby on February 22, 2007, shows a large island in the middle of one of the larger lakes imaged on Saturn's moon Titan.*

(Image from NASA/JPL)

These lakes have been discovered by radar imaging, with dark areas in the radar image indicative of smooth areas. The smooth areas are most likely liquid, rock, ice, or organic material. The detected lakes range in diameter from 3 kilometers (1.8 miles) to more than 70 kilometers (43 miles) across.

Close Encounter

Launched from Kennedy Space Center on October 15, 1997, the *Cassini-Huygens* spacecraft consists of the *Cassini* orbiter and the *Huygens* probe, which entered the atmosphere of Saturn's moon Titan and landed on its surface. *Cassini* entered Saturn's orbit on June 30, 2004. The European Space Agency's (ESA) *Huygens* probe plunged into Titan's atmosphere in January 2005. It is the first probe to land on a moon in the outer solar system. Seventeen nations contributed to building these spacecraft.

Cassini's orbital tour has provided an unprecedented wealth of images and other data, giving scientists their fullest understanding of any jovian world. The *Huygens* mission has yielded breathtaking images and other important data as it has directly sampled Titan's atmosphere. You can check on the progress of the mission and view its photos of Jupiter at www.jpl.nasa.gov.

Triton, Neptune's Large Moon

Triton's distinction among the jovian moons is a retrograde (backward) orbit—in the reverse direction of the other moons. Moreover, Triton is inclined on its axis about 20 degrees and is the only large jovian moon that doesn't orbit in the equatorial plane of its planet. Many astronomers believe that these peculiarities are the result of some violent event, perhaps a collision. Others suggest that Triton didn't form as part of the Neptunian system of moons but was captured later by the planet's gravitational field.

A Dozen More Moons in the Outer Solar System

Thanks to *Voyager*, the six medium-sized moons of Saturn have also been explored. All of these bodies are tidally locked with Saturn, their orbits synchronous, so that they show but one face to their parent planet. They are frozen worlds, made of mostly rock and water ice. The most distant from Saturn, Iapetus, orbits some 2,207,200 miles (3,560,000 km) from its parent. Because these moons orbit synchronously, astronomers speak of their leading faces and trailing faces. That one face always looks in the direction of the orbit and the other in the opposite direction has created asymmetrical surface features on some of these moons. The leading face of Iapetus, for example, is very dark in color, while the trailing face is quite light. Although some astronomers suggest that the dark material covering this moon's leading face is generated internally, others believe that Iapetus sweeps up the material it encounters in the course of its orbit.

The innermost moon of Saturn, Mimas is 115,320 miles (186,000 km) from Saturn. It is also the smallest of Saturn's moons, with a radius of just 124 miles (200 km). Mimas is very close to Saturn's rings and seems to have been battered by material associated with them. Heavily cratered overall, this small moon has one enormous crater named for the astronomer William Herschel, which makes it resemble the Empire's "Death Star" from *Star Wars.* Whatever caused this impact probably came close to shattering Mimas. Indeed, some astronomers believe that similar impacts might have created some of the debris that formed Saturn's great rings.

The medium-sized moons of Uranus are Miranda, orbiting 80,600 miles (130,000 km) above the planet; Ariel, 118,400 miles (191,000 km) out; Umbriel, 164,900 miles (266,000 km) out; Titania, 270,300 miles (436,000 km) out; and Oberon, 361,500 miles (583,000 km) out. Of these, the most remarkable is Miranda, which, in contrast to the other moons, is extremely varied geographically, with ridges, valleys, and oval-shaped faults. To the camera of *Voyager 2,* it presented a chaotic, violently fractured, cobbled-together surface unlike that of any other moon in the solar system. Clearly, this moon had a violent past, although it is unclear whether the disruptions it suffered came from within, without, or both. Some astronomers believe that Miranda was virtually shattered, its pieces coming back together in a near-jumble.

Pluto Found

Irregularities in the orbit of Uranus implied the existence of Neptune, which wasn't discovered until the middle of the nineteenth century. Yet the discovery of Neptune never did fully account for the idiosyncrasies of the Uranian orbit. Neptune also seemed to be influenced by some as-yet-unknown body. The keen, if eccentric, astronomer Percival Lowell (1855–1916) crunched the numbers for Uranus's orbit and for some 10 years searched in vain for a new planet. Clyde Tombaugh (1906–1997) was too young to have known Percival Lowell personally, but he took a job as assistant astronomer at the observatory Lowell had built in Flagstaff, Arizona. By studying photographic images, Tombaugh eventually found the planet in 1930.

That Tombaugh found Pluto a mere six degrees from where Lowell had said it would be is more a testament to serendipity than to astronomical calculations. The supposedly persistent irregularities in the orbits of Uranus and Neptune, on which he based his calculations, don't exist.

But there was Pluto nevertheless: 3.7 billion miles (5.9 billion km) from the Sun, on average, yet with an orbit so eccentric that, about every 248 years, it actually comes closer to the Sun than Neptune does.

"Dwarf Planet"

When it was first discovered, scientists considered Pluto a planet, but in the summer of 2006, the International Astronomical Union (IAU) debated the status of Pluto and decided to differentiate it from the terrestrial and jovian planets by the title "dwarf planet." It thus joins a number of other objects in this category, including Ceres (in the asteroid belt) and Eris (an object slightly larger than Pluto). As astronomers discover more objects like Pluto in the outer reaches of the solar system, they will fill out the ranks of this new celestial category.

A "New" Moon

If, having been discovered in 1930, Pluto was a late addition to our known solar system, its moon, Charon, is almost brand new, having been found in 1978. Named, fittingly, for the mythological ferryman who rowed the dead across the River Styx to the underworld ruled by Pluto, Charon is a little more than half the size of its parent: 806 miles (1,300 km) in diameter versus Pluto's 1,426 miles (2,300 km). Orbiting 12,214 miles (19,700 km) from Pluto, it takes 6.4 Earth days to make one circuit. Pluto and Charon are tidally locked—forever facing one another—the orbital period and rotation period for both synchronized at 6.4 days. Charon's status is undetermined as a satellite of a "dwarf planet." If it were orbiting the Sun (and not Pluto), it would fit the definition of a dwarf planet.

Pluto and its moon Charon were imaged with the Hubble Space Telescope (HST) in 1994. The image was taken when Pluto was 2.6 billion miles (4.4 billion km) from Earth—nearly 30 A.U. away.

(Image from R. Albrecht/ESA/NASA)

The Least You Need to Know

- Earth's gravitational field holds the Moon in orbit, but the Moon's gravitation also profoundly influences Earth, creating ocean tides.
- The Moon is biologically dead and geologically inactive.
- There is abundant evidence that the Moon was formed as a result of a collision between Earth and another planet-sized object very early in the history of the solar system.
- The jovian realm is rich in moons and planetary rings; some of these moons are volcanically active, others have measurable atmospheres.
- Io has frequent volcanic eruptions that loft material into orbit around Jupiter. Europa, also orbiting Jupiter, might have liquid water beneath its frozen, cracked surface.
- Pluto, discovered in 1930, is no longer considered a planet but is now one of a growing number of "dwarf planets" in the solar system.

Chapter 8

This World and Beyond

In This Chapter

- A theory of solar system origin and evolution
- The death of the solar system
- The search for other planetary systems
- Current and future techniques in the search

In this chapter, we explore planetary systems in a universal sense. That our solar system is just one among myriad planetary systems has become increasingly clear through the last several years. A little over a decade ago, our own planetary system was the only one we knew for sure existed. But times have changed, and the head count is up … way up. Over 200 planets outside our own solar system, or "extrasolar" planets, have been discovered. Observations since 1995 have revealed that we are not alone, at least not as a collection of planets. In fact, planets appear to be the norm around all sorts of stars.

One of the most difficult aspects about understanding the origins of our own solar system is that it has been in existence for a very long time. Based on studies of meteorites, astronomers believe the solar system is about 4.6 billion years old. The problem we face is akin to looking at a middle-aged man and being asked to describe the conditions at the moment of his birth. Where was the hospital? Who was the attending physician? We might be able to observe other births and assume that his birth was much the same, but, of course, no one living was present to watch the birth of this solar system, so uncovering its beginnings has required some serious sleuthing.

If we consider observations from our own solar system as well as those of the over 200 other planetary systems (and counting) that we are now aware of—many of them at very different states of evolution—we can better understand how planetary systems, including our own, are born, live, and die.

Solar System History

Amazingly, a few fragments from the early moments of the solar system's birth have survived to give us clues as to how the planets took shape around the youthful Sun. The most important clues to the origin of the solar system are not to be found in the Sun and planets, but in those untouched smaller fragments: the asteroids, meteoroids, and some of the planetary moons (including our own companion).

What Do We Really Know About the Solar System?

In a very real sense we have—in meteorites and moon rocks—"material witnesses" to the creation of the solar system. These geological remnants are relatively unchanged from the time the solar system was born. But how do we make up for an absence of precedents from which to draw potentially illuminating analogies?

One approach is to find another planetary system forming around a star younger than the Sun—but similar to the Sun—and draw analogies from it. The Hubble Space Telescope, among other space-borne probes, has given us tantalizing clues about the formation of planetary systems. Around the star Beta Pictoris, for example, astronomers have imaged a disk of dusty material in orbit farther from the star than the orbit of the dwarf planet Pluto is from our own Sun.

Familiar Territory

The best way to understand the formation of our solar system is to study both our own system, which is nearby, and the wide variety of other protoplanetary and planetary systems astronomers are discovering.

Let's start with what we know about the nearest planetary system, our own. The last 400 years of planetary exploration have given us these undeniable facts:

- Most of the planets in the solar system rotate on their axis in the same direction as they orbit the Sun (counterclockwise as seen from the North Pole of Earth), and their moons orbit around them in the same direction.
- The planets in the inner parts of the solar system are rocky and bunched together, and those in the outer part are gaseous and widely spaced.
- All the planets orbit the Sun in elliptical paths that are very nearly circles.
- Except for the innermost planet (Mercury), the planets orbit in approximately the same plane (near the ecliptic), and all orbit in the same direction.
- Asteroids and comets are very old and are located in particular places in the solar system. Comets are found in the Kuiper Belt and Oort Cloud, and asteroids are in the asteroid belt between Mars and Jupiter.
- In addition, it is clear that the asteroids that have been examined are some of the oldest, unchanged objects in the solar system and that comets travel in highly elliptical orbits, spending most of their time in the far reaches of the solar system.

The most important conclusion we can draw from these observations is that the solar system appears to be fundamentally orderly rather than random. It doesn't appear that the Sun formed first and then gradually captured its planets from surrounding space.

From Contraction to Condensation

Consider this possible portrait of the formation of our solar system: a cloud of interstellar dust, measuring about a light-year across, begins to contract, rotating more rapidly the more it contracts—and thereby conserving angular momentum. With the accelerating rotation comes a flattening of the cloud into a pancakelike disk, perhaps 100 A.U. across—100 times the current mean distance between Earth and the Sun.

The Birth of Planets

The original gases and dust grains that formed the nebular cloud contract into condensation nuclei, which begin to attract additional matter, forming clumps that rotate within the disk.

These clumps encounter other clumps and more matter, growing larger by accretion. Accretion is the gradual accumulation of mass and usually refers to the building up of larger masses from smaller ones through the mutual gravitational attraction of matter. That's the remarkable bottom line: our planetary system was built slowly by nothing more than gravity.

In the Orion and the Eagle Nebulae, we can see an enormous number of protoplanetary disks taking form—new worlds being born as we watch.

Close Encounter

In April 1998, astronomers working at the Keck Observatory in Mauna Kea, Hawaii, and at Cerro Tololo Inter-American Observatory in Chile reported startling new infrared and radio telescope evidence supporting the condensation theory. Studying a star known as HR4796A, 220 light-years from Earth, the astronomers discovered a vast dust disk forming around it. A doughnutlike hole, slightly larger than the distance between the Sun and Pluto, surrounds the star, and the disk itself extends more than twice the distance of the doughnut hole.

Although astronomers did not detect any planets in this very distant object, they believe the gravitational force of one or more inner planets might have caused the hole in the disk. In effect, astronomers believe they are seeing a distant planetary system in the making. At a mere 10 million years old, HR4796A is believed to be the right age for a system undergoing planetary formation.

Accretion and Fragmentation

The preplanetary clumps grew by accretion from objects that we imagine to be the size of baseballs and basketballs to planetesimals, embryonic protoplanets several hundred miles across. The early solar system must have consisted of millions of planetesimals.

Although smaller than mature planets, the planetesimals were large enough to have gravitational forces sufficiently powerful to affect each other. The result must have been a series of near misses and collisions that merged planetesimals into bigger objects but also caused fragmentation, as collisions resulted in chunks of some planetesimals being broken off. The formation of the Moon likely happened at this chaotic point in the history of the solar system.

The larger planetesimals, with their proportionately stronger gravitational fields, captured the lion's share of the fragments, growing yet larger, while the smaller planetesimals joined other planets or were "tossed out." A small number of fragments escaped capture to become asteroids and comets, the asteroids in smaller, more circular orbits and the comets in their wide-arcing elliptical paths.

Unlike the planets, whose atmospheres and internal geological activity (volcanism and tectonics) would continue to alter them, asteroids and comets remained geologically static, dead. Therefore, their material marks the date of the solar system's birth.

Astro Byte

Astronomers estimate that the evolution from a collection of planetesimals to eight protoplanets, many protomoons, and a protosolar mass at the center of it all consumed about 100 million years. After an additional billion years, scientists believe the leftover materials assumed their present orbits in the asteroid belt, the Kuiper Belt, and the Oort Cloud. The high temperatures close to the protosun drove most of the icy material into the outer solar system where (with the exception of periodic comets) it remains to this day.

An Old Family Recipe

Although substantial variety exists among the eight planets, they tend to fall into two broad categories: the large gaseous outer planets, known as the jovians, and the smaller rocky inner planets, the terrestrials (these groups we described in previous chapters).

Why this particular differentiation? As with just about any recipe in any kitchen, part of the difference is caused by heat.

Out of the Frying Pan

As the solar *nebula* contracted and flattened into its pancakelike shape, gravitational energy was released in the form of heat, increasing its temperature. Due to the inverse-square law of gravitational attraction, matter piled up mostly at the center of the collapsing cloud. Both the density of matter and the temperature were highest near the center of the system, closest to the protosun, and gradually dropped farther out into the disk.

def•i•ni•tion

The term **nebula** has several applications in astronomy, and observers often use it to describe any fuzzy patch seen in the sky. In astrophysics, a nebula is a vast cloud of dust and gas.

At the very center of the nascent solar system, where heat and density were greatest, the solar mass coalesced. In this very hot region, the carefully assembled interstellar dust was pulled apart into its constituent atoms, while the dust in the outer regions of the disk remained intact. When the gravitational collapse from a cloud to a disk was complete, the temperatures began to fall again, and new dust grains condensed out of the vaporized material toward the center of the solar system. This vaporization and recondensation process was an important step in the formation of the solar system because it chemically differentiated the dust grains that would go on to form the planets.

These grains originally had a uniform composition. In the regions nearest the protosun—where temperatures were highest—metallic grains formed because metals survived the early heat. Farther out, silicates (rocky material), which could not survive intact close to the protosun, were condensed from the vapor. Still farther out, there were water-ice grains, and, even farther, ammonia-ice grains. What is fascinating to realize is that the heat depleted the inner solar system (which is home to Earth) of water ice and organic carbon compounds. These molecules, as you will see, survived in the outer solar system and might have later rained onto the surfaces of the inner planets in the form of comets, making Earth habitable.

The composition of the surviving dust grains determined the types of planets that would form. Farthest from the Sun, the most common substances in the preplanetary dust grains were water vapor, ammonia, and methane, in addition to the elements hydrogen, helium, carbon, nitrogen, and oxygen—which were distributed throughout the solar system. The jovian planets and their moons, therefore, formed around mostly icy material. And in the cooler temperatures farthest from the protosolar mass, greater amounts of material were able to condense, so the outer planets tended to be very massive. Their mass was such that, by gravitational force, they accreted hydrogen-rich nebular gases in addition to dust grains.

Hydrogen and helium piled onto the outer planets, causing them to contract and heat up. Their central temperatures rose, but never got high enough to trigger fusion, the process that produces a star's enormous energy. Thus the jovian worlds are huge but also gaseous. They present no surface to explore.

Into the Fire

Closer to the protosun, in the hottest regions of the forming solar system, the heaviest elements, not ices and gases, survived to form the planets. Thus the terrestrial planets are rich in the elements silicon, iron, magnesium, and aluminum. The dust grains and the planetesimals from which these planets were formed were rocky and metallic rather than icy.

Close Encounter

If the central area of the forming solar system was too hot for light elements and gases to hang around, where did Earth's abundant volatile matter—the oxygen, nitrogen, water, and other elements and molecules so essential to life—come from?

One theory is that comets, which are icy fragments formed in the outer solar system, were deflected out of their orbits by the intense gravitational force of the giant jovian planets. Bombarded by icy meteoroids after the planet had coalesced and cooled, Earth was resupplied with water and other essential elements.

No matter how they got here, it is fortunate that water, ice, and organic compounds later rained down on the early Earth, or the present-day planet might be as lifeless as the Moon.

Ashes to Ashes, Dust to Dust

Just as the specifics of the formation of the solar system depended on the formation of the Sun, so its death will be intimately related to the future of our parent star. The evolution of the Sun will surely follow the same path of other stars of its size and mass, which means that the Sun will eventually consume the store of hydrogen fuel at its core. As this core fuel wanes, the Sun will start to burn fuel in its outer layers. It will grow brighter, and its outer shell will expand. It will become a red giant, with its outer layers extending as far as the orbit of Venus. When the Sun puffs up into a red giant, some 2 or 3 billion years from now, Mercury will slow in its orbit and probably fall into the Sun. Venus and Earth will certainly be transformed, their atmospheres (and, in the case of Earth, its water) being driven away by the intense heat of the swelling Sun. Venus and Earth will return to their infant state, dry and lifeless.

These changes will also have a transforming effect on the outer planets and their moons, warming them significantly. Perhaps even cold, dry Mars will become a more hospitable location than it is today. The recently explored moon of Saturn, Titan, will be warmed by the expanding Sun, perhaps causing a brief blossoming of life there.

The question remains: can we extrapolate our understanding of solar system formation to other planetary systems? Clearly some of the details must be the same, but most of the systems that we have discovered are different enough from our own to cause theorists some serious headaches.

Other Worlds: The News So Far

Until fairly recently, the existence of other planetary systems was assumed, but had not been observed. Starting in 1995, though, clear evidence for planets around other stars began to accumulate, and by June of 2007, astronomers had discovered over 200 planets orbiting other stars.

Since ancient times, humans have wondered about the possibility of other worlds, perhaps like our own. The prophet Muhammad weighed in on the topic in the seventh century C.E., commenting that many worlds with human inhabitants existed.

Christian debates about the presence of other worlds periodically raged in western Europe until 1609, when Galileo's telescope made it clear that other worlds—at least within the solar system—existed without a doubt. After other worlds were an accepted fact in the solar system, and astronomers understood that the Sun was just one of many stars, it wasn't a great leap to think that planets might exist around the other stars in the night sky.

How to Find a Planet

Models of star formation have made it clear that it would be very difficult to form a star without also forming a planet, a task akin to tossing a pizza and having it remain a tight ball of dough. But hard evidence for the existence of planets around other stars eluded astronomers until the eve of the twenty-first century. There are a number of ways to hunt for planetary systems, some more obvious than others and some more technically feasible than others. Astronomers have tried every method yet proposed, with varying degrees of success.

Take a Picture

The most obvious approach is to take a picture, which sounds simple enough. When we want a picture of a planet in the solar system, we simply point a telescope in its direction, snap a photo, and we're done. If we want a really sharp picture, we can send a probe to orbit the planet (like the *Mars Orbiter Camera* or the *Cassini-Huygens* mission to Saturn). If we want to discover planets around other stars, we can just point our telescopes in the direction of a suspicious-looking star and voila! How hard can it be?

Well, we do have a few observational problems with this simple approach, at least with our existing technology.

The first problem is the relative brightness of stars and planets. Stars shine because of the nuclear fusion occurring in their cores. Planets shine in the reflected light of their mother star. And planets are much smaller than the stars they orbit. As a result, even the largest planet in our solar system (Jupiter) is about a billion times fainter than the Sun. To put it bluntly, the "simple" imaging approach is difficult because it's very hard to see a firefly (the planet) circling a multi-megawatt searchlight (the star).

The other problem with imaging directly is that planets and stars are very close to one another relative to how far away they are from us. So our task is to look for something (a planet) very close to something that is very far away (a star). Most Earth-bound telescopes don't have enough resolving power to do this. Therefore, what would seem the simplest solution is beyond our available technology. As the resolving power of telescopes improves (both with larger diameters and adaptive optics), taking pictures will become more possible.

Watch for Wobbling

When a large planet orbits a star, the central star doesn't stay exactly fixed. It moves around a little bit, and that movement betrays the presence of an orbiting planet. The bigger the planet, the more the star wobbles.

We have two ways to detect star motion. One involves taking images that record the exact positions of the stars. The other involves observing the color spectrum of a candidate star and looking for minute changes in its color with time. Both require meticulous observations and lots of telescope time.

Not Astronomy, but Astrometry

The first technique, called *astrometry*, requires many images of a star and its nearby companions. These images are examined for minute changes in the star's position. If a particularly large planet orbits a star, it is possible to measure the wobble in the position of the star. For example, if some slimy alien astronomer were looking at the Sun, he, she, or it would see the Sun wobble back and forth every 11 or 12 years. Our alien friend would be watching the Sun's response to the orbit of Jupiter. All of the other planets would cause wobbles as well, but the wobble from Jupiter would be the dominant one.

The problem with this method is that the wobbles caused by even the largest planets are very small, at the limit of even the best telescopes now in operation. (Of course, alien technology might be far more advanced than ours.)

Do the Doppler Shift

The second technique uses the Doppler effect, or Doppler shift. You hear the Doppler effect when a fire truck races past you, the pitch of the siren going from high to low as the truck passes where you are standing. The sound is "blueshifted" (moves toward higher frequency) as the truck approaches and "redshifted" (moves toward lower frequency) as it recedes. Light from a star does the same thing.

As an orbiting planet pulls a star away from the viewer, the star's light is redshifted. When the orbiting planet pulls the star toward the viewer, its light is blueshifted. We can use the pattern of blueshifts and redshifts to determine the size and distance of the planet doing the pulling.

It should be no surprise that the first systems discovered with this technique were the ones with the biggest planets in the smallest orbits (doing the most pulling), but as more and more systems have been discovered, we are finding systems that resemble our own multiple-planet system. In 1999, the first extrasolar multiple-planet system, Upsilon Andromeda, was discovered with this technique.

Take the Planetary Transit

Solar eclipses are regular events on Earth. They occur when the Moon passes in front of the Sun, blocking nearly all of its light. Less obvious events occur when planets like Venus and Mercury pass in front of the Sun, blocking a tiny fraction of its light. These events are called planetary transits, and they happen with a fair degree of regularity because all of the planets orbit pretty much in the same plane.

However, detecting a planetary transit in a distant star takes both luck and skill. The luck comes in happening to view another planetary system "edge on." The skill comes in making very sensitive measurements of the brightness of a star. When the brightness dips in a tiny, regular, and cyclical fashion, then a transit has been discovered. The first such detection was made for the star HD 209458 in 1999. The great advantage of this sort of measurement is that the details of the "dip" can reveal the radius of the planet that is transiting, and—if a spectrometer is used—even features of the atmosphere of the orbiting planet.

Other Solar Systems: The News So Far

Astronomers were puzzled in the early days of discovering extrasolar planetary systems because almost none of the systems looked anything like our solar system. The Doppler technique turned up system after system in which a very large (Jupiterlike) planet was in a very small (Mercurylike) orbit. Debates raged about how the formation of such systems fits into the model for the formation of our own solar system. But with time, astronomers have started to discover ever smaller planets in ever larger orbits, and these systems are starting to look more and more like our own. What is interesting about this progression of discovery is that there are clearly a variety of ways in which planetary systems can form, ours being only one—perhaps not even the most common—example.

Close Encounter

The race to discover planets is heating up. In May 2007 alone, astronomers announced the discovery of 32 "new" planets orbiting nearby stars. These new discoveries have increased the number of known "exoplanets" to 242. At this rate, we will know of over 1,000 planets in just a few more years.

Don't Be So Self-Centered

Our discoveries in extrasolar planetary systems highlight the danger of coming up with a theory of planetary system formation based on one example, namely our solar system. Likely the details of the over 200 discovered planetary and protoplanetary systems will help astronomers improve upon the models that account for the formation of our own system.

Puppis: A Familiar System

In the constellation Puppis in the southern hemisphere, astronomers using the Doppler technique have detected a new planetary system very similar to our own. A planet twice as massive as Jupiter orbits the star HD70642 at 3.3 A.U. At this distance, it orbits in six years. And the star is a yellow dwarf, very much like our own Sun.

Because the newly discovered planet rounds HD 70642 in a nearly circular orbit at a distance of 3.3 A.U., the inner part of this distant planetary system is left clear for Earth-sized planets.

Perhaps more significant to the search not only for planets but also for life, circular orbits ensure that a planet will not experience temperature extremes. The Puppis system is fascinating because most of the planets discovered until now have had highly elliptical orbits, which are probably incompatible with life, at least as we know it. Perhaps Puppis contains a twin to Earth and curious living things looking back at us.

Where to Next?

A number of space missions planned in the coming decade are likely to expand the stable of known planetary systems. Thus far, all the planets discovered have been large and Jupiterlike. Although these discoveries are nothing short of stunning, there is—among some scientists and the public at large—a burning desire to discover Earthlike planets—terrestrial planets that resemble our own. Perhaps we are searching for ourselves.

Philosophical musings aside, many planned missions will pan for terrestrial gold. One mission, Terrestrial Planet Finder (TPF), has as its goal the discovery of terrestrial planets orbiting nearby Sunlike stars. After such planets are pinpointed, the TPF will perform spectroscopy, examining the atmospheres of the planets for evidence of life (the presence of oxygen, for example, or ozone, both of which would not last long in a planetary atmosphere without constant replenishment from biological sources). TPF will actually consist of two observatories, a visible light coronagraph, and a space-based optical interferometer. The optical interferometer will be used to "take a picture," finally, as we mentioned earlier. The launch of this mission is slated for the period 2012 to 2015.

Scheduled to launch even earlier, in November 2008, is the Kepler Mission, which will scan the skies for planetary transits. Its instrumentation will be able to scan thousands of stars at once and detect very faint transits, the kind that result from planets as small as Earth.

And even now, many Earth-based telescopes, including the Keck interferometer in Hawaii, are taking data in the quest for planets. The two giant, 10 m Keck telescopes are being used in conjunction to make images of the dust around nearby stars. Keck observations will also enable astronomers to measure the astrometric wobbles of stars due to Uranus-sized planets—not as small as Earth, but a major step in the right direction.

So stay tuned, we already know we have a lot of planetary company, and only time will tell what else is out there.

The Least You Need to Know

- Based on studies of meteorites, astronomers believe the solar system is 4.6 billion years old.
- The early solar system, like all young planetary systems, was filled with dust grains pulled together by gravity to form planetesimals. The composition of the dust grains, and thus the planetesimals, depended on the distance from the Sun.
- A gravitationally contracting nebula in which dust grains condense and collide to form planets can explain most of the patterns observed in our solar system and other known planetary systems.
- Stars other than the Sun are host to planetary systems, and some of these systems resemble our own.
- Three main techniques exist for discovering extrasolar planetary systems; so far, the Doppler effect technique has been the most productive. But new missions will change the landscape of discovery.

Part 3

To the Stars

By far the closest and most familiar star is the Sun, and before we leave the confines of our solar system to explore the nature of stars in general, we take a close look at our own star, its structure, and the mechanism that allows it to churn out, each and every second, the energy equivalent of 4 trillion trillion 100-watt bulbs.

From the Sun, we move on to how stars are studied and classified. We then explore how stars develop and what happens to them when they die, including very massive stars that are many times the mass of our Sun. As we probe stellar death, we'll explore the truly strange realm of neutron stars and black holes.

Chapter 9

The Sun: Our Star

In This Chapter

- The Sun: an average star
- Sunspots, prominences, and solar flares
- The layered structure of the Sun and its atmosphere
- The Sun as a nuclear fusion reactor
- Dimensions and energy output of the Sun
- Something new under the Sun: solar neutrinos

For thousands of years, star gazers had no idea that the stars, in all their beauty, had anything in common with the Sun. It's hard to see much similarity between the distant, featureless points of light against a sable sky and the great yellow disk of daytime, whose brilliance overwhelms our vision and warms our world. Yet, of course, our Sun is a star, the very center of our solar system, the parent of the terrestrial and jovian planets and their rings and moons. We are the leftovers of the formation of this burning sphere in the sky.

Part 2 of this book discussed the planets and their moons. But taken together, these objects represent only 0.1 percent of the mass of the solar system. The other 99.9 percent is found in the Sun. Ancient cultures worshipped the Sun as the source of all life. And it makes sense. The Sun is our furnace and our light bulb: the ultimate source, directly or indirectly, of most energy and light here on Earth. And because it contains almost all of the mass, it is the gravitational anchor of the solar system. Indeed, its very matter is ours. The early Sun was the hot center of a swirling disk of gas and dust from which the solar system formed some 4.6 billion years ago. If the Sun were a pie, Earth and the rest of the planets would be nothing more than a bit of flour and sugar left on the counter.

Important as it is to us, the Sun is only one star in a galaxy containing hundreds of billions of stars. Actually, astronomers feel fortunate that the Sun is so nondescript a galactic citizen. Its ordinariness lets us generalize about the many stars that lie far beyond our reach. In this chapter, we examine our own star and begin to explore how the Sun (and stars in general) generate the enormous energies that they do.

The Solar Mystery

The ancient Greek philosopher Anaximenes of Miletus believed the Sun, like other stars, was a great ball of fire. His belief was an important insight but was not entirely accurate. The Sun is not quite so simple.

In terms of human experience, the Sun is an unfailing source of energy. Where does all that energy come from? In the nineteenth century, scientists knew of two possible sources: thermal heat (like a candle burning) and gravitational energy.

The problem with thermal energy is that even the Sun doesn't have enough mass to produce energy the way a candle does—at least, not ongoing for billions of years. Calculations showed that "burning" chemically (in the manner, say, of logs in a fireplace), the mass of the Sun would last only a few thousand years. Although a Sun that was a few thousand years old might have pleased some theologians at the time, a variety of evidence showed that Earth was far older.

Astro Byte

The description of how mass has an equivalent energy is perhaps the best-known equation of all time: $E = mc^2$. E stands for energy, m for mass (in kg), and c, the speed of light (3×10^8 m/s). Do the math: a tiny bit of mass can produce an enormous amount of energy.

So scientists turned their attention to gravitational energy, that is, the conversion of gravitational energy into heat. The theory went this way: as the Sun condensed out of the solar nebula, its atoms fell inward and collided more frequently as they got more crowded. These higher velocities and collisions converted gravitational energy into heat. Gravitational energy could power the Sun's output at its current rate for about 100 million years.

But when it started to become clear that Earth was much older (early geological evidence showed that it was at least 3.5 billion years old), scientists went back to the drawing board, and the nineteenth century came to a close without an understanding of the source of energy in the Sun.

Astronomer's Notebook

Some important measurements used to express the energy output of the Sun include: watt (a measure of power, the rate at which energy is emitted by an object); solar constant (the energy each square meter of Earth's surface receives from the Sun per second: 1,400 watts); and luminosity (the total energy radiated by the Sun per second: 4×10^{26} watts).

A Special Theory

At the beginning of the twentieth century, a brilliant physicist named Albert Einstein came up with an answer. It turns out that the source of energy in the Sun is something never considered before—something that would have been called alchemy centuries earlier. It so happens that the Sun converts a tiny bit of its mass into energy through the coming together of the cores of atoms, their nuclei.

What's a Star Made Of?

The Sun is mostly hydrogen (about 73 percent of the total mass) and helium (25 percent). Other elements, found in much smaller amounts, add up to just under 2 percent of the Sun's mass. These include carbon, nitrogen, oxygen, neon, magnesium, silicon, sulfur, and iron. Over 50 other elements are found in trace amounts. Nothing is unique about the presence of these particular elements; these same ones are distributed throughout the solar system and the universe.

To produce energy, hydrogen atoms in the Sun's core plow into one another and thereby create helium atoms. In the process, a little mass is converted into energy. That little bit of energy for each collision equates to enormous amounts of energy when we count all of the collisions that occur in the core of the Sun. With this energy source, the Sun is expected to last not 1,000 years or even 100 million years, but about 8 to 10 billion years, typical for a star with the Sun's mass. The lifespan of a star, as we'll see later, depends on its exact mass at its time of birth.

How Big Is a Star?

In terms of its size, mass, and released energy, the Sun is by far the most spectacular body in the solar system. With a radius of 6.96×10^8 m, it is 100 times larger than Earth. Imagine yourself standing in a room with a golf ball. If the golf ball is Earth, a ball representing the Sun would touch the eight-foot ceiling. With a mass of 1.99×10^{30} kg, the Sun is 300,000 times more massive than Earth. And with a surface temperature of 5,780 K (compared to Earth's average 290 K surface temperature), the Sun's photosphere would melt or vaporize any matter we know.

Four Trillion Trillion Light Bulbs

Next time you're screwing in a light bulb, notice its wattage. A watt is a measure of power or how much energy is produced or consumed per second. A 100-watt bulb uses 100 joules of energy every second. For comparison, the Sun produces 4×10^{26} watts of power. That's a lot of 100-watt light bulbs—four trillion trillion of them, to be exact. This rate of energy production is called the Sun's *luminosity*. Some stars have luminosities much higher than that of the Sun, but most have lower luminosities.

The source of the Sun's power—and that of all stars, during most of their lifespan—is the fusing together of nuclei. Stars first convert hydrogen into helium, and heavier elements come later (we discuss this process in Chapter 11). The only sustained fusion reactions we have been able to produce on Earth are uncontrolled reactions known as hydrogen bombs. The destructive force of these explosions gives insight into the enormous energies released in the core of the Sun. In the core of the Sun, 600 million metric tons of hydrogen are converted into helium every second. Nuclear fusion could be used as a nearly limitless supply of energy on Earth; however, we are not yet able to create the necessary conditions, namely powerful magnetic fields, to create controlled fusion reactions.

The Atmosphere Is Lovely

The Sun doesn't have a surface as such. What we call its surface is just the layer that emits the most light. So let's begin our journey at the outer layers of the Sun, the layers we can actually see, and work our way in.

When you look up at the Sun during the day, what you are really looking at is its photosphere, the layer from which the visible photons that we see arise. The photosphere has a temperature of about 6,000 K. Lower layers are hidden behind the photosphere, and higher layers are so diffuse and faint we can see them only during total solar eclipses or with special satellite-borne instruments. Above the photosphere in the solar atmosphere are the chromosphere, the transition zone, and the corona. As we move higher in the Sun's atmosphere, the temperatures rise dramatically.

Not *That* Kind of Chrome

The part of the Sun's atmosphere called the chromosphere is normally invisible because the photosphere is far more dense and bright. However, during a total solar eclipse, which blocks light from the photosphere, the chromosphere can be seen as a pinkish aura around the solar disk. The strongest emission line in the hydrogen spectrum is red, and the predominance of hydrogen in the chromosphere imparts the pink hue.

The chromosphere is a storm-racked region, into which spicules, jets of expelled matter thousands of miles high, intrude.

Above the chromosphere is the transition zone. The temperature at the surface of the photosphere is 5,780 K, much cooler than the temperatures in the solar interior, which gets hotter closer to the core. Yet in the chromosphere, the transition zone, and into the corona, the temperature rises sharply the *farther* one goes from the surface of the Sun! At about 6,000 miles (10,000 km) above the photosphere, where the transition zone becomes the corona, temperatures exceed 1,000,000 K. (For detailed real-time views of the solar photosphere, chromosphere, and corona, see http://sohowww.estec.esa.nl.)

How do we explain this apparent paradox? We believe the interaction between the Sun's strong magnetic field and the charged particles in the corona heat it to these high temperatures and its low density makes it more difficult for it to cool because cooling happens more easily when atoms collide, which they do in high-density environments.

A Luminous Crown

Corona, Latin for "crown," describes the region beyond the transition zone which consists of highly ionized elements stripped of their electrons by the tremendous heat in the coronal region. Like the chromosphere, the corona is normally invisible, blotted out by the intense light of the photosphere. Only during total solar eclipses does the corona become visible, at times when the disk of the Moon covers the photosphere and the chromosphere. During such eclipse conditions, the significance of the Latin name becomes readily apparent: the corona appears as a luminous crown surrounding the darkened disk of the Sun. In fact, during eclipses, with the corona evident, the Sun looks most like a typical child's drawing of it.

When the Sun is active—a cycle that peaks every 11 years—its surface becomes mottled with sunspots, and great solar flares and prominences send material far above its surface.

Solar Wind: Hot and Thin

The Sun doesn't keep its energy to itself. Rather this energy flows away in the form of electromagnetic radiation and particles. The particles (mostly electrons and protons) don't move nearly as fast as the radiation, which escapes the surface of the Sun at the speed of light, but they move fast nevertheless—at more than 300 miles per second (500 km/s). This swiftly moving particle stream we call the *solar wind.* (You can keep track of the latest solar wind weather at http://umtof.umd.edu/pm/.)

The incredible temperatures in the solar corona drive the solar wind. As a result, the gases are sufficiently hot to escape the tremendous gravitational pull of the Sun. The surface of Earth is protected from this wind by its magnetosphere, the magnetic field, a kind of "cocoon," generated by the rotation of charged material in Earth's molten core. Similar fields are created around many other planets, which also have molten core material.

The magnetosphere either deflects or captures charged particles from the solar wind. Some of these particles are trapped in the Van Allen Belts, doughnut-shaped regions around Earth named after their discoverer, University of Iowa physicist James Alfred Van Allen (1914–2006). Some of the charged particles rain down on Earth's poles and collide with its atmosphere, giving rise to displays of color and light called *Auroras* (in the Northern Hemisphere, the Aurora Borealis, or Northern Lights, and in the Southern Hemisphere, the Aurora Australis, or Southern Lights). The Auroras are especially prominent when the Sun reaches its peak of activity every 11 years. The latest cycle of activity started in 1996, peaked in 2000, and ended in 2006. The next peak activity should occur in 2011.

Into the Sun

Having described the layers in the Sun's outer atmosphere, let's look at some of their more interesting aspects: the storms in the atmosphere. Solar weather in the form of sunspots, prominences, and solar flares regularly disturbs the Sun's atmosphere.

With the proper equipment—or nothing more than an Internet connection (go to http://sohowww.estec.esa.nl)—you can observe many of the signs of activity on the Sun's surface.

A Granulated Surface

The Sun's surface usually appears featureless, except, perhaps, for *sunspots*. However, viewed through a solar filter and at high-resolution, the surface of the Sun actually appears highly granulated. Now, *granule* is a relative concept when talking about a body as big as the Sun. In fact, each granule is about the size of an Earthly continent, appearing and disappearing as a hot bubble of photospheric gas rises to the solar surface.

def•i•ni•tion

Sunspots are irregularly shaped dark areas on the face of the Sun. They appear dark because they are cooler than the surrounding material. They are tied to the presence of magnetic fields at the Sun's surface.

Galileo Sees Spots

People must have seen sunspots before 1611, when Galileo—and, independently, other astronomers as well—first reported them. The largest spots are visible to the naked eye, at least when the Sun is seen through thin clouds, yet in Galileo's day, the world was reluctant to accept "imperfections" on the face of the Sun, and, as far as we know, no one studied sunspots systematically before Galileo.

From the existence and behavior of sunspots, Galileo drew a profound conclusion. He declared, in 1613, that the Sun rotated, interpreting the apparent movement of the spots across the face of the Sun as evidence of the Sun's rotation. The sunspots were just along for the ride.

Sunspots: What Are They?

Sunspots, irregularly shaped dark areas on the face of the Sun, look dark because they are cooler than the surrounding material. Strong local magnetic fields push away some of the hot ionized material rising from lower in the photosphere. A sunspot is not uniformly dark. Its center, called the umbra, is darkest and is surrounded by a lighter penumbra. If you think of a sunspot as a blemish on the face of the Sun, just remember that one such blemish might easily be the size of Earth or larger.

Sunspots sometimes persist for months and might appear singly, although, usually, they are found in pairs or groups. Such typical groupings are related to the magnetic nature of the sunspots and their correlation with the presence of magnetic "loops." Every pair of spots has a leader and a follower, with respect to the direction of the Sun's rotation, and the leader's magnetic polarity is always the opposite of the follower. That is, if the leader is a north magnetic pole, the follower will be a south magnetic pole.

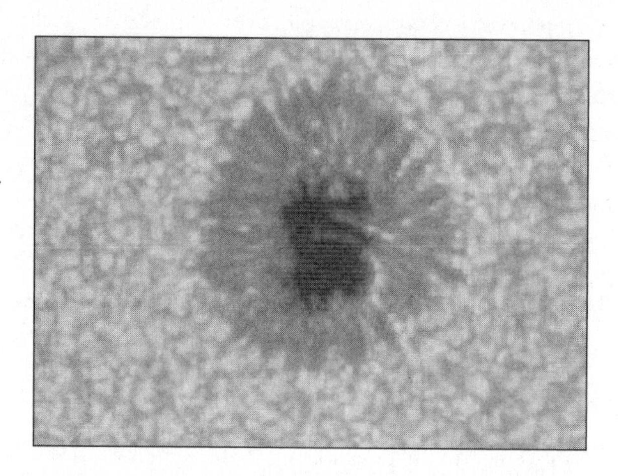

This sunspot was photographed by the National Solar Observatory at Sacramento Peak, Sunspot, New Mexico. Note the small granules and the larger sunspot umbra, surrounded by the penumbra.

(Image from NSO)

Sunspots are never seen exactly at the equator or near the solar poles, and leaders and followers in one hemisphere of the Sun are almost always opposite in polarity from those across the equator. That is, if all the leaders in the northern hemisphere are south magnetic poles, all the leaders in the southern hemisphere will be north magnetic poles.

Sunspots are thought to be associated with strong local magnetic fields. But why are the fields strong in certain regions of the photosphere?

A meteorologist from Norway, Vilhelm Bjerknes (1862–1951), concluded in 1926 that sunspots are the erupting loops of magnetic field lines, which are distorted by the Sun's differential rotation. In other words, like the giant gas jovian worlds, the Sun doesn't rotate as a single, solid unit but does so differentially, with different speeds at different latitudes. The Sun spins fastest at its equator, resulting in distortion of the solar magnetic field lines. The field lines are most distorted at the equator, so that the north-south magnetic field is turned to an east-west orientation. In places where the field is sufficiently distorted, twisted like a knot, the field becomes locally very strong, powerful enough to escape the Sun's gravitational pull. Where this happens, field lines "pop" out of the photosphere, looping through the lower solar atmosphere and forming a sunspot pair at the two places where the field lines pass through the surface into the solar interior.

Sunspot Cycles

In 1843, long before the magnetic nature of sunspots was perceived, astronomer Heinrich Schwabe announced his discovery of a solar cycle, in which the number of spots seen on the Sun reaches a maximum every 11 years (on average). In 1922, the British astronomer Annie Russel Maunder charted the latitude drift of sunspots during each solar cycle. She found that each cycle begins with the appearance of small spots in the middle latitudes of the Sun, which are followed by spots appearing progressively closer to the solar equator until the cycle reaches its maximum level of activity. After this point, the number of spots begins to decline. The most recent maximum occurred in early 2001.

Actually, the 11-year period is only half of a 22-year cycle that is more fundamental. The leading spots on one hemisphere exhibit the same polarity, meaning they are all either north magnetic poles or south—and the followers are the opposite of the leaders. At the end of the first 11 years of the cycle, polarities reverse. If the leaders had north poles in the southern hemisphere, they become, as the second half of the cycle begins, south poles.

Coronal Eruptions

Most frequently at the peak of the sunspot cycle, violent eruptions of gas are ejected from the Sun's surface. The prominences and flares can rise some 60,000 miles (100,000 km) above the photosphere and might be visible for weeks.

Solar flares are more sudden and violent events than prominences. Although they are also thought to be the result of magnetic kinks, they do not show the arcing or looping pattern characteristic of prominences. Flares are explosions of incredible power, bringing local temperatures to 100,000,000 K. Whereas prominences release their energy over days or weeks, flares explode in a flash of energy release that lasts a matter of minutes or, perhaps, hours.

The Core of the Sun

The Sun is a nuclear fusion reactor. Impressive as its periodic outbursts are, its *steady* production of energy should generate even more wonder and thought. The Sun has been churning out energy every second of every day for the last four to five billion years!

Fission Hole

On December 2, 1942, Enrico Fermi, an Italian physicist who had fled his fascist-oppressed native land for the United States, withdrew a control rod from an "atomic pile" he and his colleagues had built in a squash court beneath the stands of the University of Chicago's Stagg Field. This action initiated the world's first self-sustaining atomic chain reaction. In short, Fermi and his team had invented the nuclear reactor, and the world hasn't been the same since.

Nuclear fission is a nuclear reaction in which an atomic nucleus splits into fragments, thereby releasing energy. In a fission reactor, like Fermi's, the process of fission is controlled and self-sustaining, so that the splitting of one atom leads to the splitting of others, each fission reaction liberating more energy.

Nuclear fission is capable of liberating a great deal of energy, whether in the form of a controlled sustained chain reaction or in a single great explosion: an atomic bomb. Yet even the powerful fission process cannot account for the tremendous amount of energy the Sun generates so consistently. We must look to another process which is called *nuclear fusion.*

Whereas nuclear fission liberates energy by splitting atomic nuclei, nuclear fusion produces energy by joining them, combining light atomic nuclei into heavier ones. In the process, the combined mass of two nuclei in a third nucleus is *less* than the total mass of the original two nuclei. The mass is not simply lost but is converted into energy, a *lot* of energy.

One of the by-products of nuclear fusion reactions is a tiny neutral particle called the *neutrino.* Fermi dubbed the particle a neutrino, Italian for "little neutral one." The fusion reactions themselves produce high-energy gamma-ray radiation, but on the Sun those photons are converted into mostly visible light by the time their energy reaches the surface of the Sun. Neutrinos, with no charge to slow them down, come streaming straight out of the Sun's core.

Chain Reactions

The Sun generates energy by converting the hydrogen in its core to helium. The details are complex, so we present only the briefest overview. When temperatures and pressures are sufficiently high (temperatures of about 10 million K are required), 4 hydrogen nuclei, which are protons, positively charged particles, can combine to create the nucleus of a helium atom, 2 protons and 2 neutrons.

The mass of the helium nucleus created is slightly less than that of the four protons (hydrogen atoms) needed to create it. That small difference in mass is converted into energy in the fusion process. One of the simplest fusion reactions involves the production of deuterium (a hydrogen isotope) from a proton and a neutron. When these two particles collide with sufficient velocity, they create a deuterium nucleus (consisting of a proton and a neutron), and the excess energy is given off as a gamma-ray photon. In the Sun, this process proceeds on a massive scale, liberating the energy that lights up our daytime skies. That's a 4×10^{26} watt light bulb up there, remember?

Your Standard Solar Model

By combining theoretical modeling of the Sun's unobservable interior with observations of the energy that the Sun produces, astronomers have come to an agreement on what they call a *standard solar model*, a mathematically based picture of the interior structure of the Sun and its energy-generating "machinery." The model seeks to explain the observable properties of the Sun and also to describe properties of its unobservable interior.

Only with the solar model can we begin to describe some of the interior regions that are hidden from direct observation beneath the photosphere. Below the photosphere is the convection zone, which is some 124,000 miles (200,000 km) thick. Below this is the radiation zone, 186,000 miles (300,000 km) thick, which surrounds a core with a radius of 124,000 miles (200,000 km).

The Sun's core is tremendously dense (150,000 kg/m^3) and hot: some 15,000,000 K. We can't stick a thermometer in the Sun's core, so how do we know it's that hot? If we look at the energy emerging from the Sun's surface, we can work backward to the conditions that must prevail at the Sun's core. At this density and temperature, nuclear fusion is continuous, with particles always in violent motion. The Sun's core is a giant nuclear fusion reactor.

At the very high temperatures of the core, all matter is completely ionized—stripped of its negatively charged electrons. As a result, photons, packets of electromagnetic energy, move slowly out of the core into the next layer of the Sun's interior, the radiation zone.

Here the temperature is lower, and photons emitted from the core of the Sun interact continuously with the charged particles located there, being absorbed and reemitted. While the photons remain in the radiation zone, heating it and losing energy, some of their energy escapes into the convection zone, which boils like water on a stovetop so that hot gases rise to the photosphere and cool gases sink back into the convection zone. Convective cells get smaller and smaller, eventually becoming visible as granules at the solar surface. Thus, by convection, huge amounts of energy reach the surface of the Sun. Atoms and molecules in the Sun's photosphere absorb some of the Sun's emerging photons at particular wavelengths, giving rise to the Sun's absorption-line spectrum. Most of the radiation from a star that has the surface temperature of the Sun is emitted in the visible part of the spectrum.

The Solar Neutrino: Problem Solved

Remember that neutrinos are produced in the Sun's core as a product of the nuclear fusion reactions occurring there, a process sometimes called the proton-proton chain.

Unlike photons, which scatter in the solar interior for millions of years before getting to the solar surface, neutrinos interact only weakly with other matter and, therefore, stream directly out of the solar interior. Neutrinos, then, give us our only immediate view of events going on in the core of the Sun.

Theoretical models of the fusion reactions in the solar core predict how many neutrinos we should expect to detect, given the Sun's energy production rate. However, since the very first experiments were run to detect neutrinos from the Sun, scientists have been aware of a problem. None of the experiments detected the number of neutrinos expected based on the standard solar model. There were always too few. The discrepancy varied by experiment, but all came up short by a factor of up to ⅓.This shortfall, called the Solar Neutrino Problem (SNP), persisted for 30 years.

Later in the early 1990s, the GALLEX (GALLium EXperiment) clarified the situation somewhat by detecting lower-energy neutrinos than any other detector. These neutrinos resulted from the most common proton fusion reaction, and the results of the experiment made it clear that the shortfall in neutrinos was real, not a problem with the detectors.

So how, then, do we explain the Solar Neutrino Problem? There were two main ideas.

First: the core of the Sun might not be as hot as we thought. This would reduce the number of neutrinos produced in the core and, therefore, the number detected.

Second: neutrinos might change or "oscillate" on their journey from the Sun to Earth. There are, in fact, three different types of neutrinos (electron, muon, and tau neutrinos), like three flavors of ice cream in a restaurant.

Some recent observations at the Sudbury Neutrino Observatory (SNO), located about 1 mile (2 km) below the surface of Earth in Canada have solved the problem. The detector sits under all that rock to protect it from particles other than neutrinos. Filled with a thousand tons of "heavy" water (water containing heavy isotopes of hydrogen called deuterium), the detector can "see" flashes of light called Cherenkov radiation each time a neutrino interacts with the water. As with other neutrino detectors, the light is detected by photomultiplier tubes, which surround the tank.

A joint project of Canada, the United States, and Britain, the SNO in 2001 directly detected some of those changed neutrinos, indicating that the "oscillations" are real. A 30-year-old mystery has been solved.

The Least You Need to Know

- As staggering as the Sun's dimensions and energy output are, the Sun is no more nor less than a very average star.
- The Sun is a complex, layered object with a natural nuclear fusion reactor at its core.
- The gamma rays generated by the fusion reactions in the core of the Sun are converted to optical and infrared radiation by the time the energy emerges from the Sun's photosphere.
- Although the Sun has been a dependable source of energy for the last four billion years, its atmosphere is frequently rocked by such disturbances as sunspots, prominences, and solar flares, peaking every 11 years.
- The long-standing "Solar Neutrino Problem" has been resolved as scientists have discovered that neutrinos have a tiny amount of mass and change or "oscillate" during their journey from the core of the Sun to Earth.

Chapter 10

Giants, Dwarfs, and the Stellar Family

In This Chapter

- Nearest and farthest stars
- Observing and calculating stellar movement
- Measuring the size of a star
- Classifying stars according to stellar temperatures and chemical composition
- Determining the mass of a star
- Stellar biographies

In the great scheme of things, our own star, the Sun, is relatively accessible. We can make out detailed features on its surface and track the periodic flares and prominences that it emits. But our Sun is only 1 star of a few 100 billion in our Galaxy alone, and beyond that exist stars much farther away, parts of other distant galaxies.

Even distant Pluto is very close in comparison to our stellar neighbors. On the golf ball scale, the nearest star—which is in the Alpha Centauri system—would be 50,000 miles away. The rest of the stars in the night sky are even farther.

In this chapter, we begin to reach out to myriad cousins—quite literally distant cousins—of our Sun.

Sizing Them Up

The planets of the solar system appear to us as disks. When we know a planet's distance from us, it is simple to measure the disk and translate that figure into a real measurement of the planet's size. But train your telescope on any star, and all you see is a point of light, with no disk to measure. Pop in a higher-power eyepiece, and guess what? It's still a point of light.

Only very recently, with the advent of the Hubble Space Telescope (HST), have we been able to resolve the disks of *any* stars. In 1996, the HST took a picture of the star Betelgeuse, which was the first resolved image of a star other than the Sun. The Very Large Array, ground based in New Mexico, has also been used to image Betelgeuse. Although far away at 500 light-years, Betelgeuse is a red giant, and that means it has a very large radius indeed.

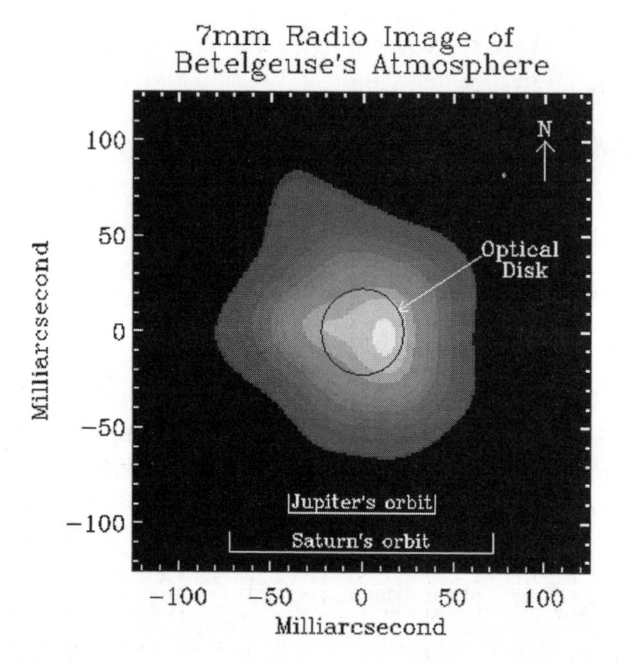

The Very Large Array in Socorro, New Mexico, recently made an image of the nearby supergiant star, Betelgeuse, the bright red star in the constellation Orion. Like the Sun, Betelgeuse appears to send plumes of gas far above its optically visible atmosphere, as seen in this radio-frequency image.

(Image from NRAO)

Radius, Luminosity, Temperature: A Key Relationship

Just because we have very few disks to measure, we don't have to give up on determining the sizes of the stars. We just have to be more clever and a bit more indirect.

First, we determine the star's temperature and mass by studying its color and visible spectrum. Then, employing numerical models of how stars hold together, we derive the quantity we are interested in—radius, for example. This indirect process is akin to looking out over a parking lot and seeing a Lexus. You might not know its size, but you know, by consulting an Internet website, that this model of Lexus is 16.4 feet long. You can clearly see that it is indeed this particular model of Lexus, so you know its length, even though you didn't actually measure it with a ruler.

A star's luminosity, its wattage or the rate at which it emits energy into space, is highly dependent on its temperature: proportional to the fourth power to the star's surface temperature. There is another important relationship for stars. A star's luminosity is not only related to its temperature but also to its surface area. Heat the head of a pin to 400 degrees F and a large metal plate to the same temperature. Which will radiate more energy? Obviously, the object with the larger surface area will. Given the same surface temperature, a larger body will always radiate more energy than a smaller one.

We can express the relationship between surface area and energy radiated in this way: a star's luminosity is proportional to the square of its radius (that's the surface area term) times its surface temperature to the fourth power: luminosity $\approx r^2 \times T^4$. So, if you know a star's luminosity and temperature, you can calculate its radius.

Let's back up a moment. How exactly do we measure a star's luminosity and temperature? Let's see.

The Parallax Principle

First, how do we know which of all the stars in the night sky are the nearest to us? For that matter, how do we know how far away any stars are? You've come a long way in this book, and on this journey we have spoken a good deal about distances—by earthly standards, often extraordinary distances. Indeed, the distances astronomers measure are so vast that they use a set of units unique to astronomy. When measuring distances on Earth, meters and kilometers are convenient units, but in the vast spaces between stars and galaxies, such units become inadequate and very clumsy. The way astronomers measure distances and the units they use depend on how far away the objects are.

We can measure distances between a given point on Earth and many objects in the solar system with radar, which can detect and track distant objects by transmitting radio waves and then receiving the returning waves the object reflects back. (Sonar is a similar technique using sound waves, and LIDAR uses light waves.) If you multiply the round-trip travel time of the outgoing signal and its incoming echo by the speed of light (which, you recall, is the speed of all electromagnetic radiation, including radio waves), you obtain a figure that is twice the distance to the target object.

Radar ranging works well with objects that return (bounce back) radio signals, but stars, including the Sun, tend to absorb rather than return electromagnetic radiation transmitted to them.

Moreover, even if you could bounce a signal off the surface of a star, most stars are so distant you would have to wait hundreds or thousands of years for the signal to make its round trip—even at the speed of light. Even the nearby (relatively speaking) Alpha Centauri system would take about eight years to detect with radar ranging, were it even possible.

So astronomers use another method to determine the distance of the stars, a method that was available long before radar was developed in the years leading up to World War II. In fact, the method is at least as old as the Greek geometer Euclid, who lived in the third century B.C.E.

This technique is called *triangulation*, and it is an indirect method of measuring distance derived by geometry using a "baseline" of known size and two angles from the baseline to the object. Triangulation does not require a right triangle, but the establishment of one 90-degree angle does make the calculation of distance a bit easier.

Astronomer's Notebook

If you have a stopwatch, you can also clap your hands and see how long it takes for the echo to return. Using the technique described for radar ranging, by multiplying the round-trip time by the speed of sound in air—about 340 m/s—you get twice the width of the canyon in meters.

It works like this. Suppose you are on one rim of the Grand Canyon and want to measure the distance from where you are standing to a campsite located on the other rim. You can't throw a tape measure across the yawning chasm, so you must measure the distance indirectly. You position yourself precisely across from the campsite, mark your position, then turn 90 degrees from the canyon and carefully pace off another point a certain distance from your original position. The distance between your two observation points is called your "baseline."

From this second position, you sight to the campsite. Whereas the angle formed by the baseline and the line of sight at your original position is 90 degrees (you arranged it to be so), the angle formed by the baseline and the line of sight at the second position will be somewhat less than 90 degrees. If you connect the campsite with Point A (your original viewpoint) and the campsite with Point B (the second viewpoint), both of which are joined by the baseline, you will have a right triangle.

Now, you can take this right triangle and, with a little work, calculate the distance across the canyon. If you simply make a drawing of your setup, taking care to draw the known angles and lengths to scale, you can measure the unknown distance across the canyon from your drawing. Or if you are good at trigonometry, you can readily use the difference between the angles at Points A and B and the length of the baseline to arrive at the distance to the remote campsite.

How Far Are the Stars?

Like the campsite separated from you by the Grand Canyon, the stars are not directly accessible to measurement. However, if you can establish two viewpoints along a baseline, you can use triangulation to measure the distance to a given star.

There is just one problem. Take a piece of paper. Draw a line 1 inch long. This line is the baseline of your triangle. Measure up from that line 1 inch, and make a point. Now connect the ends of your baseline to that point. You have a nice, normal-looking triangle. But if you place your point several feet from the baseline, and then connect the ends of the baseline to it, you will have an extremely long and skinny triangle, with angles that are very difficult to measure accurately because they will both be close to 90 degrees. If you move your point several *miles* away, and keep a 1-inch baseline, the difference in the angles at Points A and B of your baseline will be just about impossible to measure. They will both seem like right angles.

For practical purposes, a 1-inch baseline is just not long enough to measure distances of a few miles away. Now recall that if our Earth is a golf ball (about 1 inch in diameter), the nearest star, to scale, would be 50,000 miles away. So the baseline created by, say, the rotation of Earth on its axis—which would give two points 1 inch away in our model—is not nearly large enough to use triangulation to measure the distance to the nearest stars.

With the diameter of Earth a fixed quantity that is only so wide, how can we extend the baseline to a useful distance?

The solution is to use the fact that our planet not only rotates on its axis but also orbits the Sun. Observation of the target star is made, say, on February 1, and then is made again on August 1, when Earth has orbited 180 degrees from its position six months earlier. In effect, this motion creates a baseline that is 2 A.U. long—that is, twice the distance from Earth to the Sun. Observations made at these two times and places will show the target star apparently shifted relative to the even more-distant stars in the background. This shift is called *stellar parallax,* and by measuring it, we can determine the angle relative to the baseline and thereby use triangulation to calculate the star's distance.

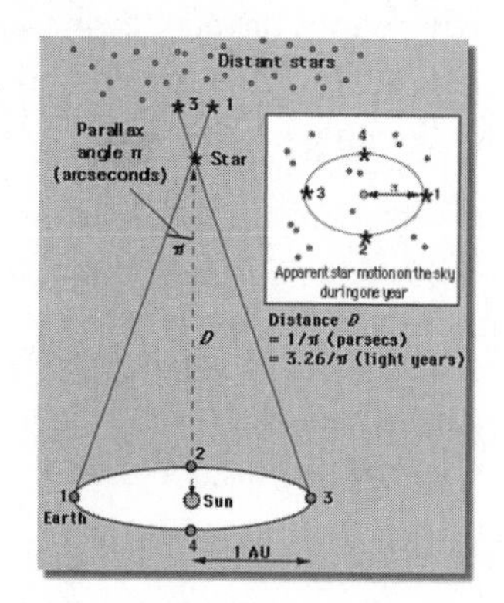

We can use Earth's orbit around the Sun to establish an enormous baseline and to measure the distances to the nearest stars.

(Image from NASA)

To get a handle on parallax, hold your index finger in front of you with your arm extended. With one eye open, line up your finger with some vertical feature, say the edge of a window. Now, keeping your finger where it is, look through the other eye. The change in viewpoint makes your finger appear to move with respect to a background object. (Warning: if you do this on the subway, people may start to move away from you.)

In our example of stellar parallax, your eyes are the two positions of Earth separated by six months, your finger is a nearby star, and the window edge is a distant background star. This method works as long as the star (your finger) is relatively close; you can experiment to see that the closer your finger is to your face, the larger its apparent motion will be. If the star is too far away, parallax is no longer effective.

Nearest and Farthest

Other than the Sun, the star closest to us is Alpha Centauri, which has the largest known stellar parallax, 0.76 arcseconds. In general, the distance to a star in parsecs (abbreviated pc) is equal to 1 divided by the stellar parallax in arcseconds—or conversely, its parallax (in arcseconds) will be equal to 1 divided by the distance in parsecs. The measured parallax, in any case, will be a very small angle (less than an arcsecond).

Recall that the full Moon takes up about 1,800" (or half a degree) on the sky, so the 0.76" parallax measured for Alpha Centauri is $^{0.76}/_{1,800}$ or less than $^{1}/_{2,000}$ the diameter of the full Moon! Using the previous rule to convert parallax into distance, we find that Alpha Centauri is $^{1}/_{0.76}$ = 1.3 pc or 4.2 light-years away. On average, stars in our Galaxy are separated by about 7 light-years. So Alpha Centauri is a bit closer than "normal." If a star were 10 pc away, it would have a parallax of $^{1}/_{10}$ or 0.1".

The farthest stellar distances that we can measure using parallax are about 100 parsecs (about 330 light-years). Stars at this distance have a parallax of $^{1}/_{100}$" or 0.01". That apparent motion is the smallest we can measure with our best telescopes. Within our own Galaxy, most stars are even farther away than this. (We'll see later that we are about 25,000 light-years from the center of our own Galaxy.) As the resolution of earthbound telescopes improves with the addition of adaptive optics, this distance limit will be pushed farther out.

Stars in Motion

The ancients thought that the stars were embedded in a distant spherical bowl and moved in unison, never changing their positions relative to one another. We know now, of course, that the daily motion of the stars is due to Earth's rotation. Yet the stars move, too. However, their great distance from us makes that movement difficult to perceive, except over very long periods of time. A jet high in the sky, for example, can appear to be moving rather slowly, yet we know that it has to be moving fast just to stay aloft and its apparent slowness is a result of its distance. A jet as far away as a star would not appear to move at all.

Astronomers think of stellar movement in three dimensions:

- The *transverse component* of motion is perpendicular to our line of sight—that is, movement across the sky. This motion can be measured directly.
- The *radial component* is stellar movement toward or away from us. This motion must be measured from the Doppler shift apparent in a star's spectrum.
- The actual motion of a star is calculated by combining the transverse and radial components, which are perpendicular to each other.

def•i•ni•tion

We determine the **proper motion** of a star by measuring the angular displacement of a target star relative to more distant background stars. Measurements are taken over long periods of time, and the result is an angular velocity (measured, for example, in arcseconds/year). If the distance to the star is known, we can convert this angular displacement into a transverse velocity.

The transverse component can be measured by carefully comparing photographs of a given piece of the sky taken at different times and measuring the angle of displacement of one star relative to background stars (in arcseconds). This stellar movement is called *proper motion*. A star's distance can be used to translate the angular proper motion into a transverse velocity in km/s. In our analogy: if you knew how far away that airplane in the sky was, you could turn its apparently slow movement into a true velocity.

Determining the radial component of a star's motion involves an entirely different process. By studying the spectrum of the target star (which shows the light emitted and absorbed by a star at particular frequencies), astronomers can calculate the star's approaching or receding velocity relative to Earth. This is the same technique used to detect extrasolar planets.

Certain elements and molecules show up in a star's spectrum as absorption lines. The frequencies of particular absorption lines are known if the source is at rest, but if the star is moving toward or away from us, the lines will get shifted. A fast-moving star's lines will be shifted more than a slow-moving one's. This phenomenon, more familiar with sound waves, is known as the Doppler effect.

Close Encounter

We have all heard of the Doppler effect. It's that change in pitch of a locomotive horn when a fast-moving train passes by. The horn doesn't actually change pitch, but the sound waves of the approaching train are made shorter by the approach of the sound source, whereas the waves of the departing train are made longer by the receding of the sound source.

Electromagnetic radiation behaves in exactly the same way. An approaching source of radiation emits shorter waves relative to the observer than a receding source. Thus the electromagnetic radiation of a source moving toward us will be blueshifted; that is, the wavelength received will be shorter than what is actually emitted. From a source moving away from us, the radiation will be redshifted; we will receive wavelengths longer than those emitted. By measuring the degree of a blueshift or redshift, astronomers can calculate the oncoming or receding velocity (the radial velocity) of a star.

How fast do stars move? And what is the fixed background against which we can measure the movement? For a car, it's easy enough to say that it's moving at 45 miles per hour relative to the road. But there are no freeways in space. Stellar speeds can be given relative to Earth, relative to the Sun, or relative to the center of the Milky Way. Astronomers always have to specify which "reference frame" they are using when they give a velocity. Stars in the solar neighborhood typically move at tens of kilometers per second relative to the Sun.

Astro Byte

Does the Sun move? Yes, the Sun and its planets are in orbit around the center of the Milky Way. The solar system orbits the Galaxy about once every 250 million years. Thus, since the Sun formed, the solar system has gone around the merry-go-round of our Galaxy 15 to 20 times.

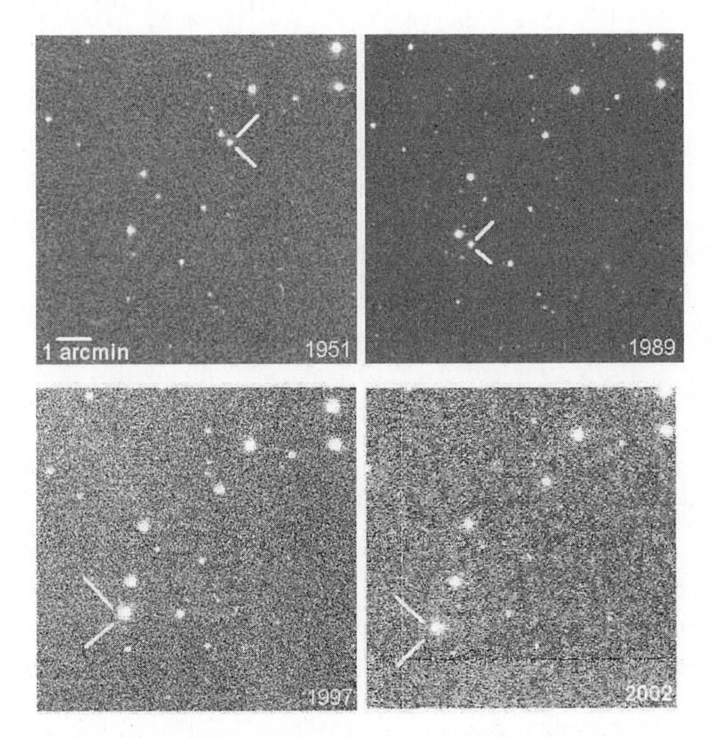

The nearby star SO025300.5+165258 has a high transverse angular motion of about 5 arcseconds per year. Bars indicate the changing position of the star with respect to more distant background stars.

(Image from B. J. Teegarden/NASA/NEAT)

How Bright Is Bright?

In ordinary English, *luminosity* and *brightness* would be nearly synonymous. But that is not so in astronomy. You are standing beside a quiet road. Your companion, a couple feet away from you, shines a flashlight in your eyes. Just then a car rounds a curve a quarter-mile away. Which is more luminous, the flashlight or the headlights? Which is brighter?

Absolutely and Apparently

Ask an astronomer which is more luminous and which is brighter, the flashlight or the headlights. She will respond that the flashlight, a few feet from your eyes, is apparently brighter than the distant headlights, but that the headlights are more luminous. Luminosity is the total energy radiated by a star each second. Luminosity is a quality intrinsic to the star, whereas brightness might or might not be intrinsic. Absolute brightness is another name for luminosity, but *apparent brightness* is the fraction of energy emitted by a star that eventually strikes some surface or detection device (including our eyes). Apparent brightness varies with distance. The farther away an object is, the lower its apparent brightness.

Simply put, a very luminous star that is very far away from Earth can appear much fainter than a less luminous star that is much closer to Earth. Or a distant, high-luminosity star might look brighter in the night sky than myriad closer, but less luminous stars. Thus, although the Sun is the brightest star in the sky, it is by no means the most luminous.

Creating a Scale of Magnitude

So astronomers have learned to be very careful when classifying stars according to apparent brightness. Classifying stars according to their magnitude (or brightness) seemed like a good idea to Hipparchus in the second century B.C.E. when he came up with a 6-degree scale, ranging from 1, the brightest stars, to 6, those just barely visible.

Unfortunately, this somewhat cumbersome and awkward system (higher numbers correspond to fainter stars, and the brightest objects—like Venus and the Sun—have negative magnitudes) has persisted to this day, although it has been expanded and refined over the years. The intervals between magnitudes have been regularized, so that a difference of 1 in magnitude corresponds to a difference of about 2.5 in brightness. Thus, a magnitude 1 star is $2.5 \times 2.5 \times 2.5 \times 2.5 \times 2.5 = 100$ times brighter than a magnitude 6 star. Because we are no longer limited to viewing the sky with our unaided eyes and larger apertures collect more light, magnitudes greater than (that is, fainter than) 6 appear on the scale. Objects brighter than the brightest stars may also be included, their magnitudes expressed as negative numbers. Thus the full Moon has a magnitude of –12.5 and the Sun, –26.8.

To make more useful comparisons between stars at varying distances, astronomers differentiate between *apparent magnitude* and *absolute magnitude*, defining the latter, by convention, as an object's apparent magnitude when it is at a distance of 10 parsecs from the observer. This convention cancels out distance as a factor in brightness, and absolute magnitude is, therefore, an intrinsic property of a star.

Some key comparative magnitudes are as follows:

- Sun: –26.8
- Full Moon: –12.5
- Venus (at its brightest): –4.4
- Vega: 0
- Deneb: 1.6
- Faintest stars visible to the naked eye: 6

Astro Byte

The Hubble Space Telescope is capable of imaging a magnitude 30 star, which has been compared to detecting a firefly at a distance equal to the diameter of Earth.

How Hot Is Hot?

A star is too distant to stick a thermometer under its tongue, but you can get a pretty good feel for a star's temperature simply by looking at its color.

The temperature of a distant object is generally measured by evaluating its apparent brightness at several frequencies in terms of a black-body curve. The wavelength of the peak intensity of the radiation emitted by the object can be used to measure the object's temperature. For example, a hot star (with a surface temperature of about 20,000 K) will peak near the ultraviolet end of the spectrum and will produce a blue visible light. At about 7,000 K, a star will look yellowish-white. A star with a surface temperature of about 6,000 K—such as our Sun—appears yellow. At temperatures as low as 4,000 K, orange predominates, and at 3,000 K, red.

So simply looking at a star's color can tell us a lot about its temperature. A star that looks blue or white has a much higher surface temperature than one that looks red or yellow.

Stellar Sorting

We can use the color of stars to separate them into rough classes, but the careful classification of stellar types didn't get under way until photographic studies of many spectra were made beginning in the early twentieth century.

Precise analysis of the absorption lines in a star's spectrum gives us information not only about the star's temperature but also about its chemical makeup. Using spectral analysis to gauge surface temperatures with precision, astronomers have developed a system of spectral classification, based on the system that astronomers at the Harvard College Observatory originally worked out. The presence or absence of certain spectral lines is tied to the temperatures at which we would expect those lines to exist. The stellar spectral classes and the rough temperature associated with the class are given in the following table.

Spectral Class	Surface Temperature
O (violet)	>28,000 K
B (blue)	10,000–28,000 K
A (blue)	7,000–10,000 K
F (blue/white)	6,000–7,000 K
G (yellow/white)	4,000–6,000 K
K (orange)	3,500–4,000 K
M (red)	<3,500 K

Stars have distinct spectra, which resemble supermarket bar codes in that they encode information about the stars from which the spectra came. In this figure, the hottest stars are at the top and the coolest stars at the bottom. Notice the star names often include HD or Henry-Draper catalog designations, from the early work done at Harvard College Observatory.

(Image from KPNO/NOAO/NSF)

The most massive stars are the hottest, so astronomers refer to the most massive stars they study as "O" and "B" stars. The least massive stars are the coolest. The letter classifications have been further refined by 10 subdivisions, with 0 (zero) the hottest in the range and 9 the coolest. Thus a B5 star is hotter than a B8, and both are hotter than any variety of A star. The Sun is a spectral type G2 star. Our Galaxy and others are chock full of type G2 and lower-temperature (less massive) stars.

From Giants to Dwarfs: Sorting the Stars by Size

We can determine the radius of a star from the luminosity of the star (which in turn can be determined if the distance is known) and its surface temperature (from its spectral type). Stars fall into several distinct classes. In sorting the stars by size, astronomers use a vocabulary that sounds as if it came from a fairytale:

- A giant is a star with a radius between 10 and 100 times that of the Sun.
- A supergiant is a star with a radius more than 100 times that of the Sun. Stars of up to 1,000 solar radii are known.
- A dwarf star has a radius similar to or smaller than the Sun.

Making the Main Sequence

Working independently, two astronomers, Ejnar Hertsprung (1873–1967) of Denmark and Henry Norris Russell (1877–1957) of the United States, studied the relationship between the luminosity of stars and their surface temperatures. Their work (Hertsprung began about 1911) was built on the classification scheme developed by Antonia Maury, a woman from the Harvard College Observatory. She first classified stars both by the lines observed and the width or shape of the lines. Her scheme was an important step toward realizing that stars of the same temperature could have different luminosity. Plotting the relationship between temperature and luminosity graphically, in what is now known as a Hertzsprung-Russell diagram or H-R diagram, these two men discovered that most stars fall into a well-defined region of the graph. That is, the hotter stars tend to be the most luminous, and the cooler stars are the least luminous.

Close Encounter

At the turn of the century, career opportunities for women were limited. However, women with specialized training in astronomy were able to find employment at the nation's observatories. The Harvard College Observatory first hired women in 1875 to undertake the daunting task of the classification of stellar types. Initially under the direction of Edward Pickering, the observatory employed 45 women over the next 42 years. Photography and the telescopes of the HCO were beginning to generate vast amounts of astronomical data—in particular, photographic plates filled with individual stellar spectra. It was more economical to pay college-educated women to pore over these data sets than to hire the same number of male astronomers.

Astro Byte

Red dwarfs are the most common stars, probably accounting for about 80 percent of the stellar population in the universe.

The region of the temperature-luminosity plot where most stars reside (indicating that they spend the majority of their lifetime there) is called the *main sequence.* Stars that are not on the main sequence are called "giants" or "dwarfs," and you will see in coming chapters how stars leave the main sequence and end up in the far corners of this temperature-luminosity graph.

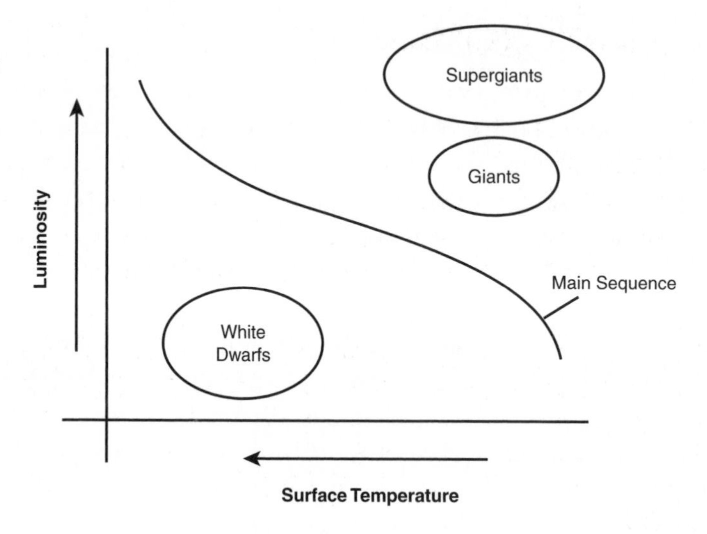

This Hertzsprung-Russell diagram, or H-R diagram, is a plot of a star's temperature and intrinsic brightness and shows where in its lifespan a particular star is. The solid line (main sequence) is where stars spend the majority of their lifetimes.

Off the Beaten Track

Although some 90 percent of stars fall into the region plotted as the main sequence, about 10 percent lie outside this range. These include *white dwarfs*, which are far less luminous than we might expect from their high surface temperatures, and *red giants*, which are far more luminous than we would expect from their relatively low surface temperatures. In coming chapters, we describe how stars leave the main sequence and end up as red giants and white dwarfs. Briefly, these stars have used up their fuel and are dying.

Stellar Mass

The overall orderliness of the main sequence suggests that the properties of stars are not random. In fact, a star's exact position on the main sequence and its evolution are functions of only two properties: composition and mass.

We can evaluate the composition if we have a spectrum of the star, its fingerprint. But how can we determine the mass of a star?

Fortunately, most stars don't travel solo, but in pairs known as binaries. (Our Sun is a notable exception to this rule.) Binary stars orbit one another.

Some binaries are clearly visible from Earth and are called *visual binaries;* others are so distant that, even with powerful telescopes, they cannot be resolved into two distinct visual objects. Nevertheless, we can observe these by noting the Doppler shifts in their spectral lines as they orbit one another. These binary systems are called *spectroscopic binaries.* Rarely, we are positioned so that the orbit of one star in the binary system periodically brings it in front of its partner. From these eclipsing binaries we can monitor the variations of light emitted from the system, thereby gathering information about orbital motion, mass, and even stellar radii.

However we observe the orbital behavior of binaries, the key pieces of information sought are orbital period (how long it takes one star to orbit the other) and the size of the orbit. After we know these factors, we can use Kepler's Third Law to calculate the combined mass of the binary system.

To the Max

The most massive star known—located in the massive star-forming region known at NGC 3603—has a mass of about 110 times that of the Sun. Part of a dense, young star cluster located a whopping 20,000 light-years from Earth, A1 is actually part of a binary system. Its companion star is also enormous, with a mass 84 times that of the Sun. There is a theoretical upper limit to how large a star can be: about 150 times the mass of the Sun. At this mass and above, gravity is unable to pull the star together.

Why is mass so important? Mass determines the fate of the star, setting the star's place along the main sequence and dictating its life span.

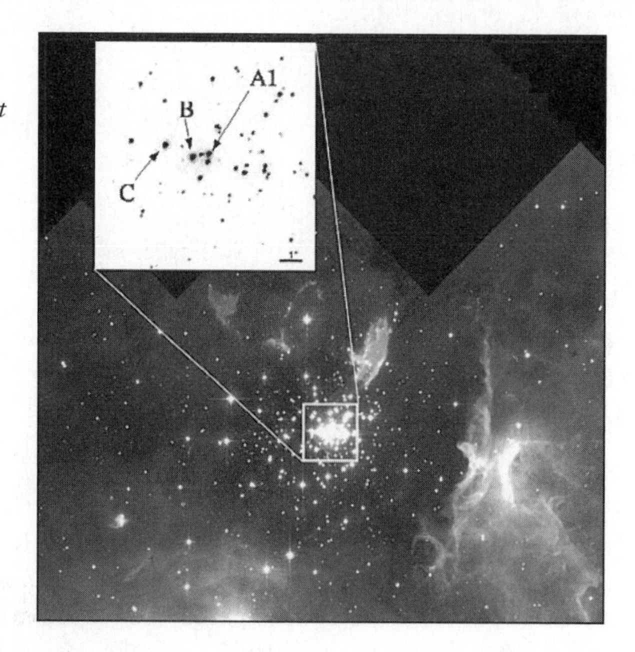

This HST image of the massive star-forming region shows the record-holding most massive star known, NGC 3603 A1.

(Image from NASA/HST)

The Life Expectancy of a Star

A star dies when it consumes its nuclear fuel, its hydrogen. One might be tempted to conclude that the greater the supply of fuel (the more massive the star), the longer it will live; however, a star's life span is also determined by how rapidly it burns its fuel. The more luminous a star, the more rapid the rate of consumption. Thus, stellar lifetime is directly proportional to stellar mass and inversely proportional to stellar luminosity (how fast it burns). Consider this analogy: a car with a large fuel tank (say a Hummer H2, which gets 13 mpg) might have a much smaller range than a car with a small fuel tank (a Toyota Celica, which gets 29 to 36 mpg). And what is the key difference? The Celica gets much better mileage and thus can go farther with the limited fuel it has.

Thus, while O- and B-type giants are 10 to 20 times more massive than our G-type Sun, their luminosity is *thousands* of times greater. Therefore, these most massive stars live much briefer lives (a few million years at best) than those with less fuel but more modest appetites for it.

A B-type star such as Rigel, 10 times more massive than the Sun and 44,000 times more luminous, will live 20×10^6 years or 20 million years. (For comparison, 65 million years ago, dinosaurs roamed Earth.) The Sun (a G2 star) might be expected to burn for $10{,}000 \times 10^6$ years (10 billion years). Our red dwarf neighbor, Proxima Centauri, an M-type star that is $\frac{1}{10}$ the mass of the Sun (and $\frac{1}{100}$ that of Rigel), is only 0.00006 times as luminous as the Sun and so will consume its modest mass at a much slower rate and might be expected to live longer than the current age of the universe.

The Least You Need to Know

- The distance to nearby stars cannot be measured directly (such as by radar ranging), but it can be determined using stellar parallax; greater distances can be determined by measuring the period of variable stars.
- Stellar motion, velocity, size, mass, temperature, and luminosity can all be measured. We can derive a rough measurement of a star's temperature from its color; hotter stars are blue, and cooler stars are red.
- By plotting the relationship between the luminosity and temperature of large numbers of stars, astronomers have noticed that most stars fall along a band in the plot called the main sequence and spend most of their lives there.
- The lifetime of a star is determined primarily by its mass; high-mass stars have short lives; low-mass stars have long lives; and all stars, including the Sun, will eventually die.

Chapter 11

The Life and Death of Stars

In This Chapter

- Star birth and the interstellar medium
- Red giants and supergiants
- From supergiant to planetary nebula
- The death of a low-mass star
- The death of a high-mass star
- Supernovae as creators of elements

John Calvin, the Protestant theologian, would have liked the theory of stellar evolution because it's all about predestination. Each and every star appears "predestined" to follow a certain path in its life, and that path is set only by the mass of the star at its birth.

However, there are a few complications. If stars have nearby companions, they can be "revived" late in life, and astronomers haven't entirely figured out the details of evolution for some types of stars.

But the mass of a star at its birth does determine where it will reside on what is called the main sequence, the period of a star's lifetime devoted to converting hydrogen into helium via nuclear fusion (see Chapter 10). The initial mass determines whether a star will be a relatively cool M-type dwarf star, a hot and massive O or B star, or something in between, such as our Sun. When the forces within a star—gravity pulling inward and the force due to the radiation pressure of fusion pushing outward—reach equilibrium, stars enter the main sequence, their relatively dull middle age. But when the forces get out of balance—as the star's core starts to run out of its primary fuel (hydrogen) and then its secondary fuel (helium)—and a star leaves the main sequence to enter its death throes, the real fun begins. The final moments of a star can be spectacular, as this chapter describes.

A Star Is Born

You probably know people with a Type A personality: they are always keyed up, hyper, super-overachievers, and you might even call them workaholics. These people often make it big, only to burn out quickly. Then there's the Type B personality: the kind of person who moves through life calmly, doing what's necessary, but no more. These people might not be very spectacular, but as with that tortoise in the fable, slow and steady can win the race.

Stellar careers show a similar range of "personalities" though the letters of the alphabet we use to label them are different. The massive, hot O and B stars are born, mature into the main sequence, only to then burn out, their hydrogen fuel exhausted, after a few million years. They are the gas-guzzlers of the galaxy. In contrast, the red dwarfs, stars of type M, low in mass and low in energy output, might not have as much fuel, but they will burn for hundreds of billions of years. Low-mass stars are the economy cars of the stellar world. They are not as flashy as their more massive cousins, but we don't see their steaming hulks by the side of the road, either.

On the Interstellar Median

Two basic components, gas and dust, fill the space between the stars. The vast majority (99 percent) of interstellar matter is what we call gas, and the majority of the gas consists of the most abundant element in the universe, hydrogen. The remainder of the material we refer to as "dust," though "smoke" might be a more accurate description. In fact, interstellar dust puts up quite a smoke screen, keeping us from viewing many regions in our own Galaxy with optical telescopes.

And the stuff between the stars is not evenly distributed. Because gas and dust have mass, the material pulls together into clouds and clumps via its own gravity. The patchy distribution of the interstellar medium means that, in some regions of the sky, astronomers can observe objects that are very distant. In other regions, where the interstellar matter is more concentrated, our optical range of vision is more limited. Think of the last time you looked out an airplane window. If the clouds were patchy, in some directions you could see the ground and in others you could not.

The lowest-density interstellar clouds consist mostly of atomic hydrogen (called HI, an H followed by a Roman numeral I and pronounced "H one"). Until the advent of radio astronomy, this atomic material was impossible to detect. Cooler, higher-density clouds contain two hydrogen atoms stuck together, called molecular hydrogen (H_2). These regions are the fuel tanks of the universe, ready to produce new stars. The regions closest to young stars are ionized, their electrons stripped away. The 10,000 K gas in these regions (called HII—"H two"—regions) emits strongly in the optical portion of the spectrum.

Astronomer's Notebook

Gas and dust are thinly distributed throughout space and are the matter from which the stars are formed. About 5 percent of our Galaxy's mass is contained in its gas and dust. The remaining 95 percent is in stars.

But let's look more closely at these different types of matter between the stars.

Blocking Light

How can mere dust block our optical view of the Milky Way? The answer has to do with the size of the dust grains. Think about this for a moment. A satellite dish needs not be solid but can be made out of a wire mesh, perforated by many small holes. This structure does not let the radio waves slip through, like water through a sieve, because radio waves are too big. They are so big, in fact, that as long as the holes are small enough, the radio waves don't even know the holes are there. They behave like pebbles too large to fall through a grate.

The radio waves reflect off the perforated surface of the satellite dish as if it were solid. All electromagnetic radiation (light included) works this way. Waves interact only with things that are about the same size as their wavelength. As luck would have it, optical wavelengths are about the same size as the diameter of a typical dust grain and are, therefore, absorbed or scattered by dust even though long-wavelength radio waves pass right through it.

The combined effect of the scattering and absorption caused by the dust produces *extinction.* That is, the dust *absorbs* blue light more than red and also *scatters* blue light more than red. As a result, the visible light that makes it through is reddened.

Interstellar reddening is increased when objects are farther away, and astronomers have to take this effect into account when determining the true color of a star.

This Hubble Space Telescope image of the nearby emission nebula and star-forming region NGC 281 was taken in October 2005. The region is about 9,500 light-years away and located in the direction of the constellation Cassiopeia. "Bok globules" are what astronomers call the clumps of gas and dust, which appear dark in the image.

(Image from NASA/HST)

You might picture interstellar dust as a kind of fog. Certainly fog, which consists of water molecules and often particulate matter, interferes with the transmission of light, as anyone who has driven in terror along a foggy mountain road can attest.

Close Encounter

If a cloud of dust is between us and a distant star, the light from the star must pass through the dust before it can get to us. Dust allows the longer (redder) wavelengths to pass, but the shorter (bluer) wavelengths are scattered and absorbed. As a result, the light that makes it through is reddened. For this reason, the setting Sun looks redder than the Sun at noon. As the sunlight passes through a thicker slab of the atmosphere near the horizon, its blue frequencies are absorbed and scattered by the atmosphere while its red frequencies pass right through.

Now, interstellar dust is so diffuse that many thousands of miles of it might not obscure anything, but stack billions upon billions of miles of material between us and a star, and it will certainly affect what we see.

Dusty Ingredients

Astronomers have been able to analyze the content of interstellar gas quite accurately by studying spectral absorption lines, the fingerprint elements create by allowing some wavelengths to pass while absorbing others. The precise composition of the interstellar dust is less well understood; however, astronomers have some clues.

The 1 percent of interstellar gas that isn't hydrogen or helium contains far less carbon, oxygen, silicon, magnesium, and iron than we would expect based on the amounts of these elements found in our solar system or in the stars themselves. We believe the interstellar dust forms out of the interstellar gas, in the process drawing off some of the heavier elements from the gas. So the dust probably contains silicon, carbon, and iron, as well as ice consisting mainly of water with traces of ammonia and methane as well as other compounds. The dust is rich in these substances, while the gas is poor in them.

Flipping Out

Most of the gas in the interstellar medium, some 90 percent, consists of the simplest element, hydrogen. Helium, the second-simplest element, accounts for another 9 percent. The remaining 1 percent consists of other elements. And the gas is mostly cold. Recall that hydrogen consists of a proton and an electron. If there is enough ambient energy, the electron gets bumped up the energy ladder into an excited state; however, most of these clouds of hydrogen are far from energy sources—stars—and emit no detectable visible light. But they do emit radio waves.

In the 1940s, Dutch astronomers were the first to appreciate that radio waves could travel unimpeded through clouds of dust and that hydrogen produces a radio-frequency spectral line. Picture both the proton and the electron as spinning like tops. They are either spinning in the same direction or in opposite directions. If they are spinning in the same direction, the hydrogen atom has a little more energy than if they were spinning in opposite directions. Every so often (it takes a few million years), an electron will spontaneously go from the high-energy state to the low-energy state. That is, it will flip its spin, and the atom will then give off a photon with a wavelength of 21 centimeters. This photon travels unimpeded through the Galaxy to our radio telescopes.

This image shows an expanding shell of neutral hydrogen in the outer Galaxy. This large shell has a diameter of 360 parsec (more than 1,000 light-years).

(Image from Canadian Galactic Plane Survey)

Because the 21 cm line (the HI line) has a particular frequency, it tells us not only *where* the gas is but also *how* it is moving. This HI line has been invaluable in mapping our own Milky Way galaxy.

Star Light, Star Bright

By definition, there is *nothing* for the amateur astronomer to see with an optical telescope in the regions of the interstellar medium that are cold. The best she can hope for is to see these regions as dark patches in the foreground of bright optical emission. Closer to stars, however, where it's hotter, the interstellar medium can be lit up to spectacular effect. Regions of gas illuminated by a nearby massive star or stars are called *emission nebulae.*

Messier objects M8 and M16 (the Lagoon Nebula and the Eagle Nebula) are two of the more famous hot clouds of interstellar matter or emission nebulae.

Emission nebulae (also sometimes called HII regions) form around young type O and type B stars. Recall that these stars are tremendously hot, emitting most of their energy in the ultraviolet portion of the spectrum. This radiation ionizes the gas surrounding the star, causing it to glow when the liberated electrons reunite (or recombine) with atomic nuclei. As the electrons cascade down the rungs of the energy ladder, they give off electromagnetic radiation at particular wavelengths. We can detect the spectral lines that result from radio frequencies through the infrared and into the optical and ultraviolet range.

Blocking Light

If there are large amounts of gas and dust between us and an emission nebula, or HII region, we won't see it optically. The dust grains will absorb or scatter the optical photons from the ionized hydrogen, and we have to resort to high-resolution infrared or radio telescopes. Toward heavily obscured regions of the Milky Way, such as the galactic center, radio and infrared observations have given us almost all the information we have.

Regions of ionized gas near young stars (HII regions) have a wide variety of sizes. Diameters range from several tens of pc for the so-called giant HII regions, all the way down to tiny HII regions (called compact or ultra-compact HII regions) with diameters as small as 1⁄100 of a pc. The smallest HII regions are deeply embedded in clouds of gas and dust, and thus are only observable at radio and infrared wavelengths.

Radio observations with the Very Large Array have revealed an enormous amount of information about the hot gas located near young, massive stars. Many of these regions are invisible optically but accessible with radio and infrared wavelength observations. This image shows the hot gas in the region that has formed many massive stars called W49A. This region is similar to the Orion Nebula, but larger and more distant, though still located in our Galaxy.

(Image from N. Homeier/C. De Pree/NRAO)

A Matter of Perspective

When we are able to see part of the interstellar medium visually, it is often because a nearby star has lit it up, but we also have other ways to see. In some directions, a cold dark cloud of gas might fall between Earth and an emission nebula. In such a case, we would see the dark cloud as a black patch against the emission from the ionized gas. In other directions, we would see small patches of the sky where there are few or no stars. It is unlikely that there are truly few stars in these directions. Almost certainly, a dense cloud of gas is in our line of sight.

The Hubble Space Telescope imaged these gas pillars in the Eagle Nebula, an emission nebula 7,000 light-years from Earth. They consist primarily of dense molecular hydrogen gas and dust. This image shows the emission from ionized hydrogen at the surface of the pillars as well as the absorption caused by the gas and dust contained in the pillars.

(Image from Jeff Hester and Paul Scowen of Arizona State University and NASA)

But we don't have to depend on luck. Even if an interstellar cloud of gas doesn't happen to fall between Earth and a background source, there are ways to detect the source. Even at the relatively cold temperatures of these clouds (100 K as compared to the nearly 10,000 K in most emission nebulae), atoms and molecules are in motion. As molecules collide with one another, they are occasionally set spinning about.

According to quantum mechanics, molecules can spin only at very particular rates, like the different speeds of an electric fan, and when a molecule goes from spinning at one rate to spinning at another, it either gives off or absorbs a small amount of energy. We can detect that energy in the form of electromagnetic radiation. For many molecules, these photons have wavelengths about a millimeter long, which we can detect by special telescopes called "millimeter telescopes" and "millimeter interferometers." With the addition of millimeter telescopes to the astronomic arsenal, no portion of the interstellar medium can escape our notice. With radio telescopes, we can image neutral (cold) hydrogen atoms. With optical, infrared, and radio telescopes, we get pictures of the hot gas near young stars. And with millimeter telescopes, we can even seek out the cold clouds of gas that contain molecular hydrogen and other molecules.

The Interstellar Medium: One Big Fuel Tank

Why so much fuss over gas and dust? Because, as far as we can tell, this is the raw material from which stars are born. Now, how does a cloud of gas become a star?

Tripping the Switch

A *giant molecular cloud* is subject to a pair of opposing forces. Gravity, as it always does, tends to pull the matter of the cloud inward, causing it to collapse and coalesce, yet as the constituent atoms of the cloud come together, they heat up, and heat tends to cause expansion. Unless some event occurs to upset the balance, the cloud will remain in equilibrium.

We're not sure what causes a molecular cloud finally to collapse and form stars, but there are several likely possibilities. The expanding shock wave of a nearby supernova explosion might be sufficient to bring about collapse, or a ripple in a galaxy disk called a density wave might also be the trigger. Some astronomers have proposed that a fast-moving star punching through a molecular cloud could prompt parts of it to collapse to form more stars.

def•i•ni•tion

Giant molecular clouds are huge collections of cold (10 K to 100 K) gas that contain mostly molecular hydrogen. These clouds also contain other molecules that we can image with radio telescopes. The cores of these clouds are often the sites of the most recent star formation.

Whatever the cause, the fact that stars exist makes it clear that some molecular clouds do become gravitationally unstable and begin to collapse. As a cloud collapses, its density and temperature increase, allowing smaller pieces of the cloud (with less mass) to collapse. When a cloud begins to collapse, it breaks into many fragments, which form scores, even hundreds (depending on the original cloud's mass) of stars of various masses. The size of each fragment determines the mass of the star that forms.

Letting It All Out

We now join a single fragment in the collapsing cloud that will become a 1-solar-mass star. Over a period of perhaps 1 million years, the cloud fragment contracts. In this process, most of the gravitational energy released by the contraction escapes into space because the contracting cloud is insufficiently dense to reabsorb the radiation. At the center of the coalescing cloud—where densities are the highest—more of the radiated energy is trapped and the temperature increases. As the cloud fragment continues to contract, photons have a harder and harder time getting out of the increasingly dense material, thereby causing the temperature at the core to rise even higher.

If the original fragment had any slight rotation, as undoubtedly it did, it will be spinning faster now. This spinning cloud contracts, like a spinning skater pulling in her arms. If the original giant molecular cloud was 10 to 100 parsecs across, then the first cloud fragments will still have been much larger than our solar system. Depending on the eventual stellar type, a cloud fragment on the verge of becoming a protostar might be somewhat smaller than our solar system. This stage of star formation is typically accompanied by dramatic jets of out-flowing material, called Herbig-Haro objects.

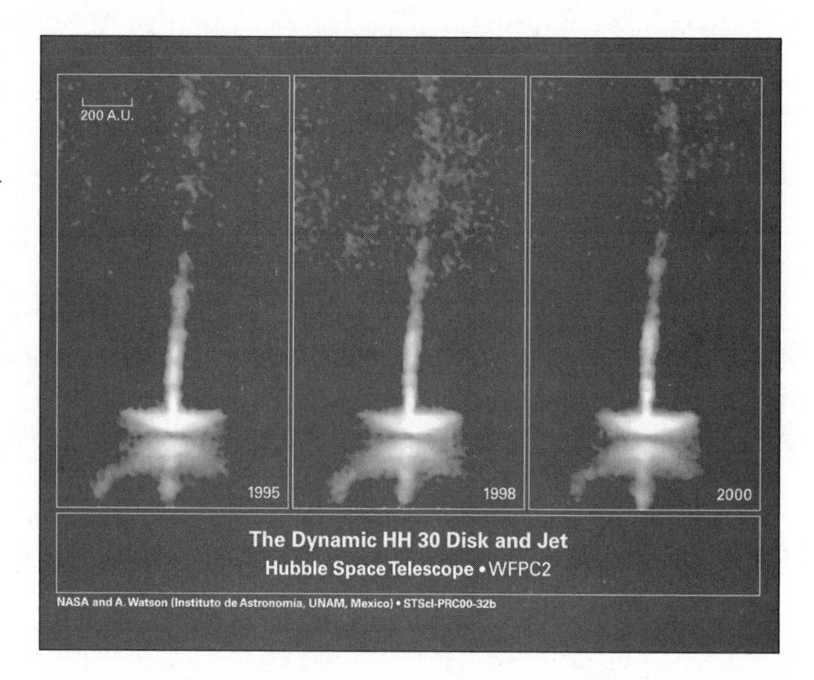

Herbig-Haro objects are jet-like structures that appear to arise during the process of low-mass star formation. High-mass stars might experience very brief, scaled-up versions of these impressive outflows.

(Image from NASA, Alan Watson, Universidad Nacional Autonoma de Mexico, Mexico et al.)

Not Quite a Star

The evolution of a protostar is characterized by a dramatic increase in temperature, especially at the core of the protostar. Still too cool to trigger nuclear fusion reactions, its core reaches a temperature of 1 million K. The protostar is also still very large, about 100 times larger than the Sun. Although its surface temperature at this stage is only half that of the Sun, its area is so much larger that it is about 1,000 times more luminous. At this stage, it has the luminosity and radius of a red giant. A solar mass star will not look this big and bright again until it is on its deathbed, about 10 billion years in the future.

Despite the tremendous heat produced by its continuing collapse, which exerts an outward-directed force on the protostar, the inward-directed gravitational force is still greater.

Through the course of some 10 million years, the protostar's core temperature increases fivefold, from 1 million K to 5 million K, while its density greatly increases and its diameter shrinks, from 100 times to 10 times that of the Sun.

Despite the increase in temperature, the protostar becomes less luminous because its surface area is getting much smaller now. Contraction continues but slows as the protostar approaches equilibrium between the inward-directed force of gravity and the outward-directed force of radiation pressure. The star now nears the part of its life called the main sequence.

When its core temperature reaches about 10 million K, the star begins to fuse hydrogen into helium. The force of gravity is balanced by the pressure of the fusion-produced heat.

A Collapsed Soufflé

Not all cloud fragments become stars. If a fragment lacks sufficient mass, it will still contract, but its core temperature will never rise sufficiently to ignite nuclear fusion. Failed stars are known as brown dwarfs.

In the Delivery Room

Analyzing the process of human gestation and birth is relatively easy. Little theorizing is required because all one needs is about nine months of free time to make some direct observations.

The birth of a low-mass star, however, may consume 40 to 50 million years—obviously more time than any observer can spare. High-mass stars, as we have said, have shorter lives and more spectacular deaths. They also appear to collapse and fall onto the main sequence more rapidly than low-mass stars. In fact, they live their entire lives in less time than it takes a single low-mass star to form.

See You on the Main Sequence

After they have matured and assumed their rightful places on the main sequence, stars, regardless of their mass, enjoy a relatively long, stable period in their lives. Indeed, stars in this phase do only one thing in their incredibly hot cores: fuse hydrogen into helium, producing great amounts of energy exactly as we described for the Sun in Chapter 9. This core hydrogen burning keeps a star alive or, more precisely, it maintains a star's equilibrium, the balance between the radiation pressure sustained by fusion pushing out and the force of gravity pulling in.

Running on Empty

From our human perspective, the main sequence life span of an average G-type star such as the Sun seems like an eternity. A good 10 billion years goes by before this kind of star enters the late stages of its life, having fused a substantial amount of the hydrogen in its core into helium.

At this point, the delicate equilibrium between gravity and radiation pressure shifts, and the structure of the star begins to change. At the core of the star is a growing amount of helium "ash." We refer to it as ash—a misnomer, really—because it is the end product of the "burning" or fusion of hydrogen. Although hydrogen is fusing in the core, the more massive helium that settles to the star's center cannot reach a sufficient temperature to fuse into a heavier element. As the helium ash dilutes the supply of hydrogen in the core, the force of gravity starts to win out over the pressure of the slowing fusion reactions. As a result, the core begins to shrink.

Astronomer's Notebook

When a main sequence star becomes a red giant, it swells up to many times its original radius and becomes far more luminous. When the Sun swells up as a red giant, its luminosity will be about 2,000 times greater than at present, and its radius will be about 150 times greater. That means that the Sun's outer layers will reach out to about 0.7 A.U. (about the orbit of Venus). Earth is at 1 A.U. When this phase happens, Earth's oceans and atmosphere will evaporate away, leaving only the planet's original rocky surface. But don't lose any sleep—those end times are two to five billion years away.

The More Things Change ...

These changes in the core beget other changes farther out. The process of shrinking releases gravitational energy throughout the star's interior. This energy release increases the core temperature, allowing hydrogen located outside the star's core to begin fusing while the helium core continues to contract and heat up. At this point, the star's situation is rather paradoxical. Its outer layers swell up and shine brightly (called *shell hydrogen burning*), while its core, filling with helium, continues to collapse. It is a short-lived state.

The transition from a G-type star (example: our Sun) on the main sequence to the next stage of its career, a red giant, consumes only 100 million years. That's a lot of time for a human, but it's a blink of an eye relative to the 10-billion-year lifetime of a G-type star: only 1 percent of its total main sequence lifetime.

A Giant Is Born

The helium-enriched core of the aging star shrinks. As it does, the gravitational energy released raises the temperature in the hydrogen-burning shell, increasing the tempo of fusion. With this increase, more and more energy dumps into the outer layers of the star, increasing its temperature—and its pressure. So while the core shrinks, the outer layers of the star expand dramatically, cooling as they do. As this happens, the star becomes more luminous (because it's larger) and cooler, moving to the upper right-hand side of the H-R diagram. The resulting star is a called a red giant. For a solar mass main sequence star, a red giant will be about 100 times larger than the Sun, with a core that is less than 1 percent of the size of the star.

Close Encounter

The night sky offers many examples of red giants. Two of the most impressive are Aldebaran, the brightest star in the constellation Taurus, and Arcturus, in Boötes. Look to the constellation Orion for an example of a red supergiant, Betelgeuse (pronounced *Beetlejuice* by some).

A Flash in the Pan

A red giant continues on its unstable career for a few hundred million years, outwardly expanding and inwardly shrinking. At some point, however, the shrinking of the core raises its temperature sufficiently to ignite the helium that has been patiently accumulating there. It's like the piston in a car engine on the compression stroke. When the temperature gets high enough, the combustion reaction is triggered. Only, in a star, it's a nuclear fusion reaction, not chemical combustion, and its initiation requires temperatures in the millions of degrees. Now, along a timeline measured in tens of millions of years, something very sudden occurs. In a process that consumes only a few hours, not millions or billions of years, helium starts to burn in an explosion of activity called the *helium flash*—the explosive onset of helium burning in the core of a red giant star. The helium now fuses into carbon, and the star settles into another, much shorter, equilibrium.

After the helium flash, the helium in the core rapidly fuses into carbon and oxygen. The star's outer layers shrink, but they become hotter, so the star gets bluer and less luminous. Once the helium is exhausted in the star's core, it's nearly the end of the road for a low-mass star. Such stars don't have enough mass to raise the core temperature sufficiently to fuse carbon into heavier elements.

Red Giant *Déjà-Vu*

In astronomical terms, the equilibrium that results from helium core burning doesn't last long, only some tens of millions of years. Fusion proceeds rapidly, as the star goes about converting helium into carbon and some oxygen. But there's much less helium than there was hydrogen in the star, so this fuel runs out much sooner.

Now something very much like the process that created the first red giant phase replays, only with helium and carbon instead of hydrogen and helium. As helium is used up in the core of the star, the carbon ash settles and the core shrinks yet again, releasing heat. This triggers helium-burning in one shell, which, in turn, is enveloped by a hydrogen-burning shell, which is left over from the last time this happened. The heat generated in these burning shells causes the star's outer layers to swell, and the star becomes a red giant once again. And when stars return to the red giant phase a second time, they are generally bigger and more luminous than the first time.

We're Losing It

A star might last in this second red giant phase for a mere 100,000 years before its carbon core shrinks to an incredibly dense inert mass about 1/1,000 the current radius of the Sun—or about the size of Earth. The surrounding shells continue to fuse carbon and helium, and the outermost layers of the star continue to expand and cool.

The outer layers of the star are now so far from the core that the star's gravity can no longer hold them, and they are able to lift off and move out into interstellar space, often in several distinct shells. These outer layers slough off from the star like spherical smoke rings, leaving behind the bare, hot, carbon-rich core of the star. The cast-off outer layers of the star, which can contain 10 to 20 percent of the star's mass, are misleadingly called planetary nebulae. They are the ejected gaseous envelopes of red giant stars, shells of gas that are subsequently illuminated by the ultraviolet photons escaped from the hot, white dwarf star that remains. Despite their name, they have nothing whatsoever to do with planets.

We Prefer to Be Called "Little Stars"

The carbon-rich core of the illuminating star, at the center of the planetary nebula, continues to glow white hot, with surface temperatures of about 100,000 K—much hotter, for example, than the Sun. Nuclear fusion in the star has now ceased. So what keeps it from collapsing further? What is balancing gravity in these stars?

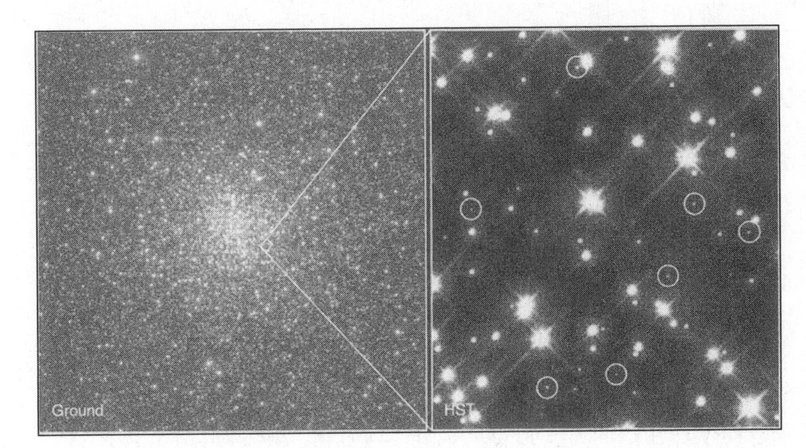

This Hubble Space Telescope image (right) reveals a population of faint white dwarfs (the circled stars) in globular cluster M4 (Messier object number 4), in the direction of Scorpius. The panel on the left is a view of this region of the sky from a ground-based telescope.

(Image from Harvey Richer of the University of British Columbia and NASA)

The white-hot core is so dense that the electrons themselves, which usually have plenty of room to move around freely, are close to occupying the same position with the same velocity at the same time. The laws of nature forbid electrons to actually do this, so the gas, therefore, can contract no further. A gas in this situation is called a *degenerate electron gas*, and the pressure of degenerate electrons is all that holds the star up against collapse at this point. The great astrophysicist Subrahmanyan Chandrasekhar (for whom the Chandra X-ray Observatory is named) first showed that stars could support themselves from further collapse in this way. This white-hot core, about the size of Earth but much more massive—about 50 percent as massive as the Sun—is called a white dwarf. One teaspoon of its carbon core would weigh a ton on Earth.

As the white dwarf continues to cool, radiating its stored energy, it changes color from white to yellow to red. Astronomers theorize that, when it ultimately has no more heat to radiate, it will become a black dwarf, a dead, inert ember, saved from gravitational collapse by the resistance of electrons to being compressed beyond a certain point. Some astronomers have proposed that the carbon atoms will eventually assume a lattice structure. The stellar corpse might become, in effect, an enormous diamond with half the mass of the Sun.

What's Nova?

Nova, the Latin word for "new," seemed to early astronomers an appropriate term for stars that suddenly appeared in the sky. Obviously, they were *new* stars. Well, they were new in that they hadn't been observed before. But, in fact, a nova is a phenomenon associated with a very *old* star.

When a binary pair of stars forms, it is unlikely the two will have the same mass. And because the mass of a star determines its lifetime, the stars in a binary pair will (more often than not) be at different points in their evolution. Sometimes, when a white dwarf is part of a binary system in which the companion star is still youthful (on the main sequence or in a red giant phase), the dwarf's gravitational field will pull hydrogen and helium from the outer layers of its companion. This gas accumulates on the white dwarf, becoming hotter and denser until it reaches a temperature sufficient to ignite the fusion of hydrogen into helium. The flare-up is brief—a matter of days, weeks, or months—but, to Earthly observers, spectacular.

Given the right circumstances, a white dwarf in a binary pair can go nova repeatedly, reigniting each time enough material is stolen from its companion star.

The Life and Death of a High-Mass Star

Up to this point, we've concentrated on the life of a star like the Sun. In truth, most of what we have described, up to the planetary nebula phase, also applies to a high-mass star. Stars of significantly lower mass than our Sun enter the main sequence and stay there, not for millions or billions of years, but much longer. Stars of significantly greater mass than the Sun (somewhere between 5 and 10 solar masses) have a very different, and more dramatic, destiny.

Fusion Beyond Carbon

Astronomers generally think of 5 to 10 solar masses as the dividing line between low- and high-mass stars. That is, stars about 5 to 10 times more massive than the Sun die in a way very different from stars of lesser mass. The major difference in their evolution is that high-mass stars are able to fuse not only hydrogen, helium, carbon, and oxygen but also heavier elements in their cores. As core burning of one element finishes, it begins burning in a shell outside the core, resulting in a nested series of shells by the time a high-mass star reaches the end of its life.

Last Stop: Iron

Hydrogen fusion produces helium ash that settles to the star's core. Then helium fusion produces carbon ash, which again settles to the star's core as if the star were some enormous centrifuge.

As the fusion of each heavier element proceeds, the core layers progressively contract, producing higher and higher temperatures. Unlike low-mass stars, in a high-mass star the gravitational forces are sufficient to drive core temperatures high enough to fuse carbon into oxygen, oxygen into neon, neon into magnesium, and magnesium into silicon. The end of the road, however, is iron. When a massive star has iron building up in its core, the grand finale is near. The reason is that for every element up until iron, energy was released when nuclei were fused. But to fuse iron into heavier elements *requires* energy. In terms of fusion, therefore, iron is a dead end. To get beyond iron, the universe needs a large input of energy. Because the periodic table of the elements does not stop with iron, you can guess that high-mass stars can somehow provide the required energy.

The evolution of the high-mass star is rapid. Its hydrogen burns for 1 to 10 million years, its helium for less than 1 million years, its core of carbon a mere 1,000 years, oxygen for no more than a year, and the fusion of silicon consumes only a week. An iron core grows as a result, but for less than a day. Just before its spectacular death, a massive star consists of nested shells of heavier elements within lighter elements, all the way down to its iron core.

Over the Edge

At this point, the core of a high-mass star is, in effect, an iron-rich white dwarf supported by its degenerate electrons. But there is a problem. The mass of a high-mass stellar remnant is so large that gravity overwhelms even the resistance of electrons to having the same position and velocity. A mass of 1.4 solar masses is sufficient to overwhelm those electrons, which are fused with protons to create neutrons. The temperatures in the core of the star become so high that all the work of fusion is rapidly undone. The iron nuclei are split into their component protons and neutrons in a process called photodisintegration.

As the core of the star collapses under its own gravity, the electrons combine with protons to become neutrons and neutrinos. The neutrinos, largely unimpeded, escape into space, the heralds of impending disaster. These neutrinos can (and have) been detected by the same neutrino detectors used to study the Sun.

The core of the star only stops its collapse when the entire stellar core has the density of an atomic nucleus. This sudden halt in the collapse causes a shock wave to move through the outer layers of the star and violently blow off its outer layers.

Supernova: So Long, and Thanks for All the Fusion

"Violently" is an understatement. Whereas the process of evolution from a hydrogen-burning star to a collapsing core has consumed 1 to 10 million years, the final collapse of a high-mass core takes less than a second and will end in a *core-collapse supernova.*

def•i•ni•tion

Core-collapse supernova is the extraordinarily energetic explosion that results when the core of a high-mass star collapses under its own gravity.

You Know the Type

We recognize two types of supernovae. Type I supernovae contain little hydrogen, whereas Type II are rich in that element. Only Type II supernovae are associated with the core collapse of high-mass stars. Type I supernovae are associated with our friends the white dwarfs.

Supernovae as Engines of Creation

As you might expect, an explosion as tremendous as that of a supernova creates a great deal of debris. The Crab Nebula, in the constellation Taurus, is the remnant of a supernova that appeared in 1054 C.E. Chinese astronomers left records of that event, reporting a star so brilliant that it was visible for a month in broad daylight. The bright radio source Cassiopeia A is also a supernova remnant.

Close Encounter

No supernova has appeared in our Galaxy, the Milky Way, since 1604. Theory predicts a supernova occurrence in our Galaxy every 100 years or so. We are, therefore, a bit overdue and might be in for a spectacular display any day now. The cosmic rays and electromagnetic radiation that would rain down on Earth if a nearby Type II supernova were to go off (say within 30 to 50 light-years) could have results that would make a massive asteroid impact look like fun. Fortunately, no stars that close to us are massive enough to generate a Type II supernova.

Amateur astronomers have been credited with many supernova discoveries. For an impressive example of supernova discoveries by Puckett Observatory, see www.astronomyatlanta.com/nova.html.

But supernovae create more than glowing remnants. Hydrogen and helium, the two most basic elements in the universe, are also the most primitive, having existed before the creation of the stars. A few other elements, carbon, oxygen, neon, silicon, and sulfur, are created by nuclear fusion in low- and high-mass stars, but many of the elements critical to life in the universe are created only in supernova explosions. Only in these explosions is there enough energy to bring nuclei together with sufficient force to create elements heavier than iron.

The only elements that existed at the beginning of the universe were hydrogen and a little bit of helium, beryllium, and lithium. Stars generated the rest of the periodic table. Each one of us, therefore, contains the debris of a supernova explosion.

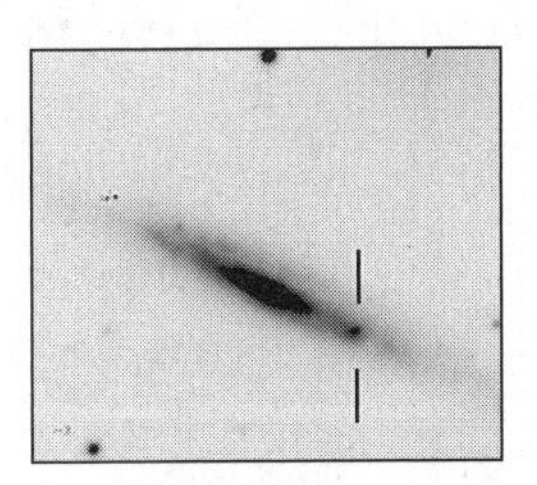

Discovery *image of a Type II Supernova 2003J. Horizontal bars indicate position of the supernova.*

(Image from T. Puckett and M. Peoples)

Leftovers

What could possibly survive a supernova? A supernova explosion pushes things out, away from the star's core, but, depending on the original mass of the star, something will be left behind: either a bizarre object known as a neutron star or, even stranger, a black hole.

And You Thought Your Roommate Was Dense

Although the explosive shock wave originates in the core, it doesn't start at the core's very center. If the core's mass was between 1.4 and 3 solar masses, the remnant at the center will be a ball of neutrons known as a neutron star.

In size, this so-called neutron star is small by astronomical standards—and tiny, compared to the high-mass star (many times the radius of our Sun) of which it was a part. The neutron star's diameter would be something over 12 miles (20 km) and its density a staggering 10^{17} kg/m^3. All of humanity (if compressed to the density of a neutron star) would be the size of a pea.

Are the Stars Spinning?

Neutron stars don't stand still. Just like the stars of which they are remnants, they rotate. And because they have collapsed from a much larger size, they spin very rapidly. Remember, thanks to conservation of angular momentum, a rotating body spins faster as it shrinks, like a whirling skater drawing in her arms. A neutron star has shrunk from a body hundreds of times larger than the Sun to one that is smaller than Earth. Earth takes 24 hours to make one revolution. A very massive but very small neutron star rotates in a fraction of a second.

The rate of rotation isn't the only property that intensifies in the neutron star. Its magnetic field is many times stronger than that of the parent star because the lines of the magnetic field are compressed along with the matter of the core itself. The combination of rapid rotation and a powerful magnetic field serves to announce the presence of some neutron stars in the universe.

Astro Byte

When pulsars were first detected in the late 1960s, their signals were so regular that some astronomers thought they might be a sign of extraterrestrial intelligence. However, the large number of pulsars detected and the neutron star theory of pulsars provided an adequate and simpler explanation.

In the late 1960s, S. Jocelyn Bell Burnell was a graduate student at Cambridge University working with Anthony Hewish looking for interesting sources of radio emission. They detected one very strange signal: a short burst of radio emission followed by a brief pause and then another pulse. The pulses and pauses alternated with great precision—as it turns out, with a precision greater than that of the most advanced and accurate timepieces in the world.

Neutron stars that form from very massive stars (perhaps 30 to 40 solar masses), can create *magnetars*, objects that generate the most powerful magnetic fields we know, thousands of times more powerful than those around a normal pulsar. Magnetars flash X-rays and, occasionally, gamma rays as a result of their more powerful magnetic fields.

In 1974, Hewish alone received the Nobel Prize in Physics for the discovery of the radio signals now called *pulsars*.

A Stellar Lighthouse

Imagine the pulsar as a stellar lighthouse. At the magnetic poles of the neutron star, though not necessarily aligned with the star's rotational axis, are regions in which charged particles are accelerated by the star's magnetic field. These regions, which rotate with the star, radiate intense energy. As the neutron star rotates, a beam of electromagnetic radiation, especially intense in the radio regime, sweeps a path through space. If Earth lies in that path, we detect the pulsar.

Thus, all pulsars are neutron stars, but not all neutron stars will necessarily be pulsars, at least not from our vantage point. If the beam of a particular neutron star doesn't sweep past Earth, we will not detect its radio pulsations. Pulsar periods can range anywhere from milliseconds to a few seconds.

Morbid Obesity

In the next chapter, we explore the nature of an even stranger supernova leftover: a black hole. If the core of the star is more massive than three solar masses, not even neutron degeneracy pressure—the fact that the neutrons cannot be compressed further—can support it. In this case, the gravitational collapse continues with nothing to halt it, and a black hole is born.

The Least You Need to Know

- A star's mass is the primary determinant of the course its life will take: high-mass stars are short-lived (tens of millions of years), while low-mass stars are long-lived (tens of billions of years).
- When a star of any mass has fused most of the hydrogen in its core, its days are numbered.

- Low-mass stars evolve into red giants and, ultimately, into white dwarfs and planetary nebulae; although some white dwarfs in binary systems can periodically reignite as novae, white dwarfs gradually cool, finally becoming burned-out embers of carbon and oxygen.
- High-mass stars (stars greater than 5 to 10 solar masses) die spectacularly when their cores collapse, creating a supernova whose remnant is either a neutron star or a black hole.
- Many of the elements in the periodic table that are important to life on our planet are produced only in supernova explosions.

Chapter 12

Black Holes: One-Way Tickets to Nowhere

In This Chapter

- Definition of a black hole
- Relativity theory
- "Seeing" black holes
- Recent black hole discoveries

Astronomy requires us to contemplate distance scales and time spans far beyond our everyday experience. Can we really fathom our Sun swelling up to the size of the orbit of Venus or the power of a supernova explosion, briefly bright enough to outshine its host galaxy? Or even the billions of years for which our star, the Sun, has been faithfully pumping out energy?

If you thought understanding the lives of stars made you stretch your mind, get ready for even more strenuous mental calisthenics because we have saved some of the strangest, least intuitive objects in the universe for this chapter. So now let's explore one of the end states of a massive star: a black hole.

When the iron core of a massive star is collapsing, it might stop when the entire core of the star has the density of an atomic nucleus, making it a neutron star. If the core is massive enough—more than 3 solar masses—the collapse becomes unstoppable, and the result is a black hole.

Under Pressure

In the case of ordinary stars, equilibrium is reached when the outward-directed forces of radiation pressure, derived from fusion reactions, are in balance with the inward-directed forces of gravity. Neutron stars, however, produce no new energy; instead, they radiate away the heat that is stored in them. They resist the crush of gravity not with the countervailing radiation pressure from fusion but with neutrons so densely packed they simply cannot be squeezed any more. Astronomers call these *degenerate neutrons.* They are like passengers in a Tokyo subway car: jammed in by truncheon-toting subway monitors until there is absolutely no room for more.

If this is the case, the neutron star isn't so much in equilibrium as it is in stasis. A stalemate exists between the irresistible force of gravity and immovable objects in the form of a supremely dense ball of neutrons.

But if the force of gravity is large enough, the collapse is apparently unstoppable, and not even neutron degeneracy can save the star's corpse.

The Livin' End

Incredible though it seems, if a star is massive enough, it will continue to collapse on itself.

Forever.

It's as if those Tokyo subway passengers all suddenly fall into a point in the center of the car and keep falling. Forever.

Remember a white dwarf evolves from a low-mass parent star (a star less than 5 to 10 solar masses) and the resulting white dwarf can be no more massive than 1.4 solar masses. If it has a higher mass, gravity will overwhelm the tightly packed degenerate electron pressure, and the core will continue to collapse. When a star's mass is greater than 1.4 solar masses, its core collapse continues, and it will blow off its outer layers as a supernova. If the mass of the remaining core is greater than 1.4 but less than 3 solar masses, the remnant will be a neutron star. However, the "specs" for a neutron star also have an upper mass limit. Astronomers believe a neutron star can be no more massive than about three times the mass of the Sun. Beyond this point, its core of neutrons will yield to gravity's pull.

What results from this apparent stalemate between the force of gravity and the incompressibility of neutrons? Do we just get an even smaller and denser neutron star?

No, not at all. We get something completely different, an object from which there is literally no escape.

When an extremely massive star is ripped apart in a supernova explosion, it can produce a supernova remnant so massive that the subsequent core collapse cannot be stopped. When this happens, *nothing* escapes the attractive forces near the core—not even electromagnetic radiation, including visible light photons. This is a black hole, so called because not even light can get out.

Astronomer's Notebook

The parent star of a collapsing neutron star core that is more than 3 solar masses would have to be 20 to 30 times more massive than the Sun. In other words, only main sequence stars with masses upwards of 20 solar masses will ever collapse into black holes. Which stars are they? They are a small subset of the massive, hot O and B stars and are far more rare than stars like the Sun.

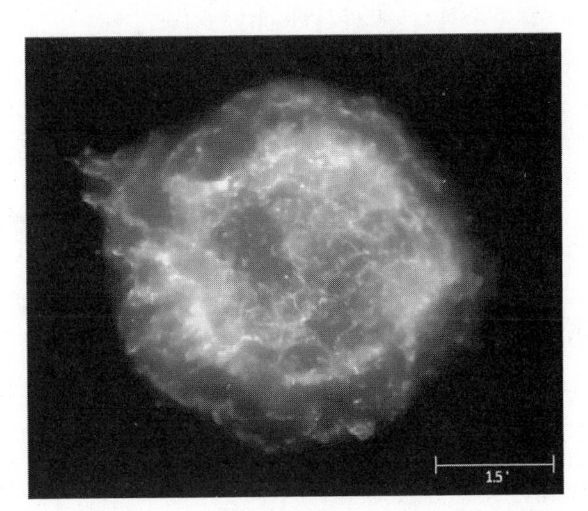

This x-ray image is of the supernova remnant in Cassiopeia. The exploded remnants of dying massive stars persist for centuries.

(Image from Chandra X-Ray Observatory)

No Escape

Although light—and thus information—cannot escape from a black hole, the black hole has certainly not ceased to exist. It is still a physical object with mass. That is the reason it creates a gravitational field, just as Earth or any other object with mass does. But how do we talk about the size of a black hole when we have just described it collapsing without end? Is it infinitely small?

Well, in fact, it does collapse to a point called a "singularity"; however, a dimension is also associated with black holes. We can talk about the mass of a black hole as well as its radius, but it's a very different kind of radius, as you'll see in a moment. First, let's take a quick detour.

Building a rocket capable of escaping Earth's gravitational pull requires an engine capable of delivering sufficient thrust to achieve a velocity of about 7 miles per second (11 km/s). This *escape velocity* depends upon two factors: the mass of the planet and its radius. For a fixed-mass object, the smaller the radius, the greater the escape velocity required to get free of its gravitational field.

def•i•ni•tion

Escape velocity is the velocity necessary for one object to escape the gravitational pull of another. The larger the mass (and smaller the distance), the greater the required escape velocity.

So as the core of a star collapses—with its mass remaining constant—the escape velocity from its surface increases rapidly. You might now wonder: What happens when a star has no more "surface" that we can talk about? What is the surface of a black hole? And is there a limit to the increase in this escape velocity?

Nature has one very strict speed limit: the speed of light. Nothing in the universe, not even photons carrying information from distant reaches of the universe, can move faster than the speed of light: 984,000,000 feet per second (300,000,000 m/s). This is the upper limit to escape velocity. When a body of a given mass reaches a certain—very small—size, objects would have to move faster than the speed of light to escape. Because nature does not allow this, escape is not an option.

What's That on the Horizon?

But how can we talk about the "surface" of a black hole? A black hole has collapsed to a point of infinite density, and a point, by definition, has no surface.

The German astronomer Karl Schwarzschild (1873–1916) first calculated what we now call the Schwarzschild radius of a black hole. For a star of a given mass, this value is the radius at which escape velocity would equal the speed of light (and, therefore, the radius within which escape is impossible). For Earth, the Schwarzschild radius is the size of a marble, about 0.4 inches or 1 centimeter. For a neutron star at 3 solar masses, the Schwarzschild radius is 5.58 miles (9 km). So this radius doesn't define a literal "surface" so much as a characteristic property of a black hole.

Remember that the collapse of a black hole is in some sense infinite. Our theoretical 3-or-more solar-mass stellar core will not stop shrinking just because it has reached the Schwarzschild radius. It keeps collapsing. When it is smaller than the Schwarzschild radius, however, it will effectively disappear. Its electromagnetic radiation (and the information that it carries) is unable to escape.

Astro Byte

How "big" would an Earth-mass black hole be? If the Earth were compressed to the size of a marble—yet retained its current mass—a velocity greater than the speed of light would be required to escape its marble-sized surface. Such an object would be, effectively, a black hole. Black holes are "black" because nothing, not even light, can escape their gravitational pull.

We spoke earlier about electromagnetic radiation carrying energy and information out into the universe. Because we cannot receive radiation from within the Schwarzschild radius, we cannot get any information from there, either. Events that occur within that radius are hidden from our view. For this reason, the Schwarzschild radius is also called the *event horizon*. As with any horizon, we cannot see past it.

Relativity

A full understanding of black holes and the phenomena associated with them requires knowledge of Albert Einstein's theory of general relativity. Einstein's most famous works are his two theories relating time and space: special relativity and general relativity. Special relativity deals with the ultimate speed limit, the speed of light, whereas general relativity is a theory of gravity. General relativity gives a more complete description of gravity's effects than Isaac Newton's eighteenth-century description, and it can explain some anomalies that Newtonian mechanics cannot. It's not that Newtonian mechanics is wrong, but since Einstein, it's just considered a special case of the more all-encompassing general relativity. Newtonian mechanics applies as long as the masses and velocities are not extraordinary. But in the environment of black holes, things become extraordinary indeed.

Curved Space Ahead

Whereas Newton introduced the concept of gravitational force as a property of all matter possessing mass, Einstein proposed that matter does not merely attract matter, but rather it warps the space around it. For Newton, the trajectory of an orbiting planet is curved because it is subject to the gravitational influence of, for example, the Sun. For Einstein, the planet's trajectory is curved because space itself has been curved by the presence of the massive Sun. This change in view represents a fundamental shift in the way we think about the universe. One major difference between Newton's and Einstein's theories of gravitation is that if mass distorts space, then massless photons of light should feel the effect of gravity, just as matter does.

Like any good theory, Einstein's ideas addressed questions that had gone unanswered and made testable predictions. In two particular examples, his theory explained some tiny but persistent peculiarities in the orbit of Mercury and successfully predicted that the Sun's mass was sufficient to bend light rays passing very close to it.

Albert's Dimple

One way to imagine space in Einstein's view is as a vast rubber sheet with a heavy bowling ball creating a big dimple in it. The mass of the bowling ball distorts the sheet, which is a two-dimensional representation of space. In this model, a massive object distorts space itself.

Now, the sheet is two-dimensional. But space is three-dimensional. We can't picture that distortion applied to our three-dimensional universe, but Einstein called that distortion "gravity."

Imagine that instead of making a dimple in the sheet (as the bowling ball does), an object were to make an infinitely deep sinkhole. That region of space would be a black hole.

In the Neighborhood

We have stated that no radiation can escape from within the Schwarzschild radius. But what of the region just outside it? It turns out that material near the black hole does produce observable radiation. Matter that strays close enough to the event horizon to be drawn into it does not remain intact, but is stretched and torn apart by enormous tidal forces—the same forces that cause the tides on Earth, but much stronger. In the process, energy can be released in the form of x-rays. Bright x-ray sources are beacons that might point the way to black holes nearby.

Here's a Thought (Experiment)

No one could ever visit a black hole and live to tell about it. A spaceship, let alone a human being, would be torn to pieces by tidal forces as it approached the event horizon.

Faced with situations impractical or impossible to observe directly or to test physically, scientists typically construct thought experiments, methodical exercises of the imagination based on whatever data are available.

Postcards from the Edge

So here's a thought experiment: suppose it were possible to send an indestructible probe to the event horizon of a black hole.

Next, suppose we equipped the probe with a transmitter broadcasting electromagnetic radiation of a known frequency. As the probe neared the event horizon, we would begin to detect longer and longer wavelengths. This shifting in wavelength is known as a gravitational redshift, and it occurs because the photons emitted by our transmitter lose some energy in their escape from the strong gravitational field near the event horizon. The reduced energy would result in a reduction in frequency (and, therefore, a wavelength increase) of the broadcast signals—that is, a redshift.

The closer our probe came to the event horizon, the greater the redshift. At the event horizon itself, the broadcast wavelength would lengthen to infinity, each photon having used all of its energy in a vain attempt to climb over the event horizon.

Suppose we also equipped the probe with a large digital clock that ticked away the seconds and was somehow also visible to us. (Remember, this is a *thought* experiment.) Through a phenomenon first described in Einstein's special relativity called time dilation, the clock would appear (from our perspective) to slow until it actually reached the event horizon, whereupon it—and time itself—would appear to slow to a crawl and stop. Eternity would seem to exist at the event horizon, and the process of falling into the black hole would appear to take forever.

As the wavelength of the broadcast is stretched to infinite lengths, so the time between passing wave crests becomes infinitely long. Realize, however, that if you could somehow survive aboard the space probe and were observing from inside rather than from a distance, you would perceive no changes in the wavelength of electromagnetic radiation or in the passage of time. Relative to you, nothing strange would be happening. Moreover, as long as the physical survival of your craft in the enormous tidal forces of the black hole was not an issue, you would have no trouble passing beyond the event horizon. But to remote observers, you would have stalled out at the edge of eternity. It's all a matter of point of view.

Into the Abyss

What's inside a black hole? We have no idea. Not because we're not curious, mind you, but because we can literally get no information from beyond the event horizon.

Theoreticians refer to the infinitely dense result of limitless collapse as a *singularity*.

We do not yet have a description of what happens to matter under the extreme conditions inside a black hole. Newton and Einstein moved our knowledge of the universe ahead in two great leaps. An understanding of these strange objects might be the next great leap forward in our comprehension of the universe.

The Latest Evidence

Can we actually *detect* a black hole? Can we see a black hole at all?

Not directly, but we can certainly see one's effects. A black hole is like some horror movie monster. We might never see the monster itself, but evidence of its existence is everywhere: its footprints, claw marks on the trees, and the muddy trail that leads back to the swamp. So what are the latest muddy trails that lead us to black holes?

Wouldn't X-Rays Kill a Swan?

In our own Galaxy, a bright source of x-rays is found in the constellation Cygnus (the Swan) known as Cygnus X-1, located at a distance of some 8,000 light-years. It is the binary companion of a B star, and the x-rays from it flare up and fade quickly, indicating that it is very small in radius. Remember we mentioned that x-rays are often emitted in the neighborhood of a stellar-remnant black hole. In addition, the x-ray source has no visible radiation and a mass (inferred from the orbit of its companion) of about 10 solar masses. In this case circumstantial evidence might be enough to convict.

Black Holes in Our Own Backyard

The universe is a big place, so having a black hole in our own Galaxy means that it is, relatively speaking, quite close. In fact, astronomers have detected a black hole and an old star that are orbiting one another only 5,000 light-years away from us. Poetically named GRO J1655-40, the black hole is an object referred to as a *microquasar*—a black hole that has the mass of a star and acts like a small version of the black holes that can have more than a million times the mass and are at the cores of active galaxies called *quasars.*

The black hole is apparently the remnant of a supernova explosion that the companion star survived, and the companion orbits the black hole every 2.6 days. What is perhaps most interesting about this pair of stars is that they are tearing through the Galaxy at high velocity—about four times the velocity that is normal for a star in this part of the Galaxy.

Now *That's* a Black Hole

The black hole in Cygnus is what we call a stellar remnant black hole. In theory, black holes can have much higher masses—masses far greater than a stellar core. These are called supermassive black holes and are found in the cores of galaxies. The Hubble Space Telescope image shown on the next page is about as close as we have gotten to a supermassive black hole. Released on May 25, 1994, the image shows a whirlpool of hot (10,000 K) gas swirling at the center of an elliptical galaxy (see Chapter 14) known as M87, 50 million light-years from us in the direction of the constellation Virgo.

Using the Hubble's Faint Object Spectrograph, astronomers Holland Ford and Richard Harms were able to measure how light from the gas is redshifted and blueshifted as one side of the 60-light-year-radius disk of gas spins toward us and the other away from us. The radius at which gas is spinning and the velocity of its rotation tell us how much matter must be within that radius. With the high resolution of the Hubble Space Telescope and ground-based radio telescopes like the Very Long Baseline Array (VLBA), astronomers can trace the rotation of gas to smaller and smaller distances from the center. Doing so, they find that even at very small radii, the gas is rotating at velocities that indicate something very massive still lies within that radius.

Black holes are almost certainly responsible for another observed galactic phenomenon. Radio astronomers have identified what they call "radio galaxies," which make themselves known by enormous radio jets that can extend from them for hundreds of thousands of light-years. The sources of these jets are black holes. Now, when two galaxies collide and the black holes merge, the direction of the jet emitted can abruptly change, creating what is known as an X-type radio source. Thus, it appears that X marks the spot—at least for some black holes.

Finally, recent observations have also suggested that, when galaxies collide, the black holes present at the centers of the two galaxies merge as well, creating the mother of all mergers and acquisitions.

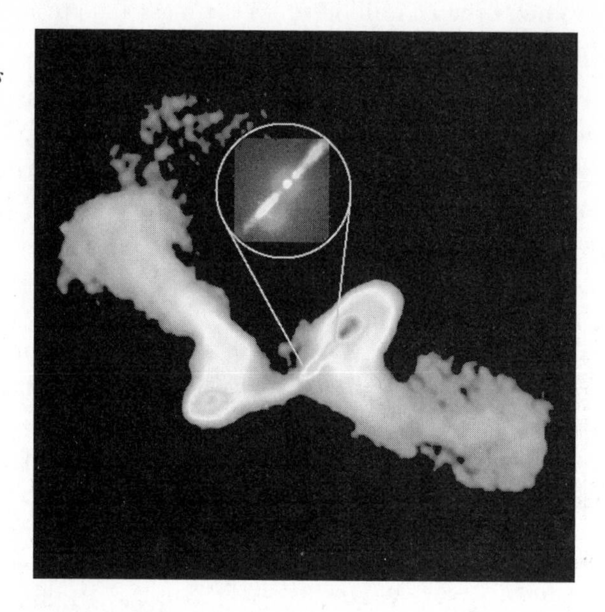

This image shows the radio jets of the source NGC 326 as imaged with the VLA. Inset is the Hubble Space Telescope image of the same galaxy.

(Image from NRAO/AUI, STScI)

The Least You Need to Know

- A neutron star is one possible remnant of a massive star that has exploded as a supernova.
- Black holes are the other possible remnant of collapsed massive star cores; if the core has a mass greater than three solar masses, the collapse cannot stop and a black hole is born.
- If you don't get too close to a black hole, its effects (in terms of gravity) are no different than a star of the same mass; they are *not* giant galactic "vacuum cleaners."
- According to general relativity, black holes, stars, and anything with mass distorts the space around it, and this distortion can influence the path of both particles *and* light.

Part 4

Way Out of This World

Now we turn from individual stars and stellar phenomena to some of the largest structures in the universe: the galaxies. We begin at home, with a chapter on our Galaxy, the Milky Way. Then we move on to an explanation of the three major galaxy types—spiral, elliptical, and irregular—and explain how galaxies, themselves vast, are often grouped into even larger galaxy clusters, which, in turn, may be members of superclusters. Here is cosmic structure on the greatest scale.

In addition to "normal" galaxies, we'll discuss another galaxy type, the "active galaxy," which emits large amounts of energy. We explore objects such as Seyfert galaxies and radio galaxies, as well as one of the most spectacular cosmic dynamos in the universe, quasars. We also consider the nature of a phenomenon that appears to be occurring in distant galaxies: gamma ray bursters (GRB).

Chapter 13

The Milky Way: Our Very Own Galaxy

In This Chapter

- The Milky Way's birth, evolution, size, and motion
- Our place in the Milky Way
- The structure of the Milky Way
- A variety of variable stars
- The mystery of dark matter
- The galactic center: a black hole?

Ancient societies were well aware of the Milky Way, a fuzzy, wispy, whitish band that arced across the sky; however, they had no idea that their world was a tiny part of this very arc. Without the aid of a telescope, one can't see that this band actually consists of billions of individual stars. For this reason, ancient cultures described the Milky Way variously: as a bridge across the sky, as a river, as spilled cornmeal from a sack dragged by a dog, or as the backbone of the heavens. So imagine Galileo's shock when he first looked through his telescope at the Milky Way and the fuzziness resolved into an enormous number of individual stars. Why, he must have asked, were those stars all arrayed across a bridge on the sky? Several centuries would pass before we determined our place in that grand arc.

With many of us living in light-polluted cities and suburbs, the Milky Way can be difficult or impossible to see. Under the best viewing conditions, however, it is a stunning sight, a majestic band of light sometimes extending high above the horizon, depending on your location and the time of year.

When we look at it, what we see is our own Galaxy from the inside. For the Milky Way is our home, so let's take a closer look at our place in the Milky Way and how it might have come to be.

Where Is the Center and Where Are We?

Galileo and other astronomers soon realized that the stars around us were not randomly distributed, but in a distinct way, confined to a narrow band of the sky. In the late eighteenth century, William Herschel proposed the first model of our Galaxy, suggesting that it was disk-shaped and that the solar system (centered on the Sun) lay near the center of this disk. This model held some psychological comfort (Earth having been so recently elbowed from the center of the universe by Copernicus and Galileo), but the model also had a problem. Herschel failed to account for what we now know as interstellar extinction because he assumed that space was essentially transparent. But the disk of our Galaxy is "foggy" with gas and dust, and we can see only so far into the "fog."

As a result, we only *appear* to be at the center of the disk as determined from star counts in various directions. Actually we are not in the center because the dust in our Galaxy absorbs visible light so that we can only see out into the disk for a limited distance—the same limited distance in all directions along the arc. We clearly run out of stars in two directions when we look "up" and "down" out of the disk of our Galaxy.

Not until early in the twentieth century did astronomers realize we were not, in fact, at the center of our own Galaxy, but at its outer suburbs.

Home Sweet Galaxy

The universe is not evenly populated with stars and other objects. Just as large distances exist between individual stars, there are large distances between the collections of stars called galaxies. Galaxies (of which the Milky Way is one) are enormous collections of stars, gas, and dust.

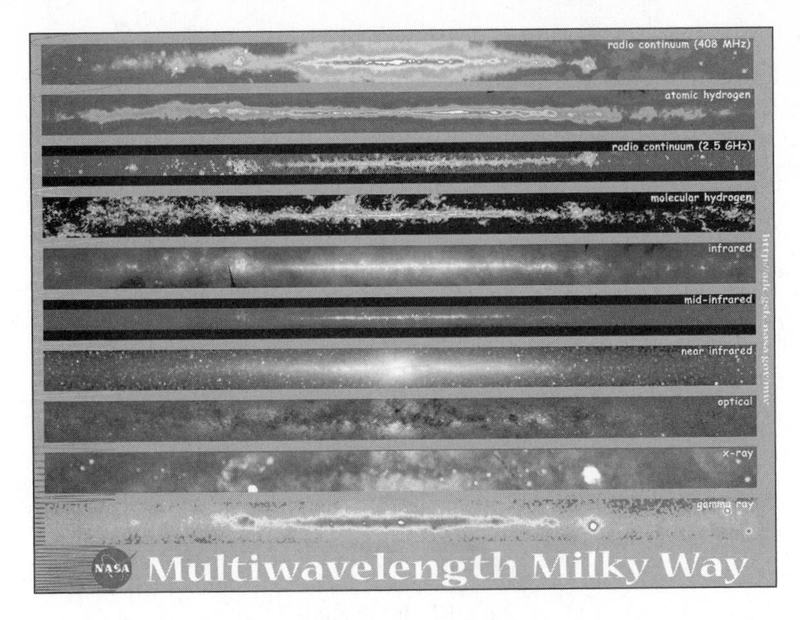

This image shows the Milky Way Galaxy as imaged at a variety of wavelengths. The center of the Galaxy is in the center of each image. The optical image (labeled) is the most familiar view, but other wavelengths give us a unique view of our home Galaxy.

(Image from NASA)

Typical galaxies contain several hundred billion stars. About 100 times as many stars exist in our Galaxy as there are people on Earth. Astronomers often refer to our own Galaxy with a capital "G" to distinguish it from all of the other galaxies in the universe.

A Thumbnail Sketch

Because we live within the Milky Way, we see it in profile. The Milky Way, viewed at this angle, is shaped rather like a flying saucer—that is, it resembles a disk that bulges toward its center and thins out at its periphery. The thickest part we call the bulge, and the thin part, the disk. (Astronomers aren't *always* in love with obtuse jargon!) Later you will see how we figured this all out, but we know now that the center of our own Galaxy is in the *Galactic bulge*, in the direction of the constellation Sagittarius.

def•i•ni•tion

The **Galactic bulge** or nuclear bulge is a swelling at the center of our Galaxy. The bulge consists of old stars and extends out a few thousand light-years from the Galactic center.

Our solar system is situated in part of the thinned-out area, the so-called Galactic disk. Our location in the Galaxy explains what we see when we look at the Milky Way on a clear country summer night, far from city lights and smog. The wispy band of light arcing across the sky is our view *into* this disk. The band of light is the merged glow of the stars close enough for us to see optically—so many that, to the unaided eye, they are not differentiated as separate points of light. A dark band of obscuration runs the length of the arc. The presence of large amounts of dust in the disk blocks this light.

When we look away from the arc in the sky, we don't see much of the Milky Way because we are looking out of the disk. Looking into the disk is like looking at the horizon on Earth: lots of things block our view, including houses, trees, and cars. But if we look straight up into the sky, we can see much farther—we might even see a plane flying overhead. When we look more or less perpendicular to the disk of the Galaxy, we can see much farther; this is the direction in which to look for other galaxies, for example.

While looking in this direction—up in the air, as it were— astronomers discovered another component of the Galaxy, the *globular clusters.* These are collections of several hundred thousand mostly older stars, held together by their mutual gravitational attraction. They are generally found well above and below the disk of the Galaxy. Reasoning that globular clusters should gather around the gravitational center of the Galaxy, Harlow Shapley in the early twentieth century used the distances and positions of these collections of stars to determine where in the Galaxy we are located.

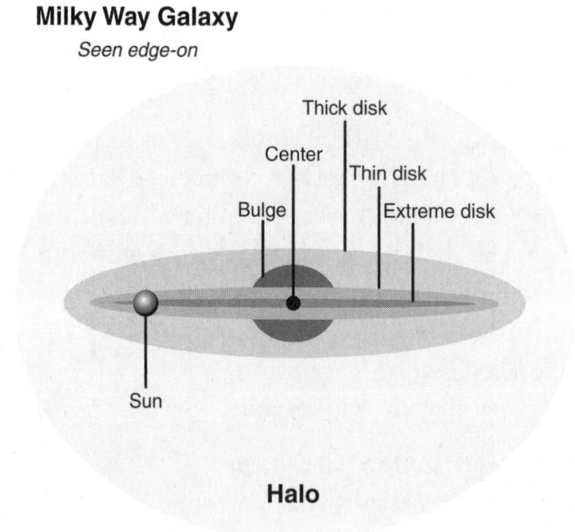

This simple diagram of our view of the Milky Way shows the principal parts. Studies have shown that the Milky Way is an ordinary spiral galaxy.

Compare and Contrast

None of us will live long enough to watch a single star evolve. At our radius, the Milky Way takes about 225 million years to rotate just once! So to overcome the limitation of the human life span we chart the development of a certain mass star by observing many similar stars in various stages of development. The same is true for galaxies. We can sometimes look for galaxies similar to our own in order to gain the necessary perspective and make generalizations about the Milky Way.

The Andromeda galaxy (Messier Object 31) is over 2 million light-years from us and is the only object outside our Galaxy that is visible to the naked eye. Like the Milky Way, Andromeda is what we call a spiral galaxy, with distinct, bright, curved structures called spiral arms, which reveal where the young, massive stars in the galaxy are forming. The universe contains many spiral galaxies that astronomers can observe to study aspects of the Milky Way that are hidden by our location in it.

Let's Take a Picture

While we can see Andromeda with our eyes or a modest amateur telescope as a fuzzy patch of light, we require photographic equipment or an electronic CCD camera to make out the kind of detail in a galaxy that we are used to seeing in glossy astronomy magazine pictures. To see sweeping arms and dark bands of dust in other galaxies, we need to collect more light, either by using a larger aperture (a big telescope) or waiting longer (taking a sufficiently exposed conventional or electronic photograph).

Measuring the Milky Way

One way to gauge the size of the Milky Way is to look at other, similar galaxies, such as Andromeda. If we know how far away such a galaxy is and can measure its angular size on the sky, we can calculate how big it is.

But determining the distance of Andromeda and other galaxies can be a serious challenge. Parallax as a distance indicator is out because we know that the apparent angular shift is equal to 1 divided by the distance in parsecs. Stars farther than 100 pc are too distant to use the parallax method, and those stars are still in our own Galaxy. We would require half a microarcsecond resolution (0.0000005") to see parallax of even the nearest galaxy, Andromeda, and no existing telescopes have resolution approaching that.

So what are we to do? In 1908, Henrietta Swan Leavitt, working at the time under the direction of Edward Pickering at the Harvard College Observatory, discovered one very good distance indicator.

def•i•ni•tion

A **variable star** is a star that periodically changes in brightness. A **cataclysmic variable** is a star, such as a nova, that changes in brightness suddenly and dramatically as a result of interaction with a binary companion star. An **intrinsic variable** changes brightness because of rapid changes in its diameter. **Pulsating variables** are intrinsic variables that vary in brightness in a fixed period or span of time.

Over centuries of star gazing, astronomers had noted many stars whose luminosity was variable—sometimes brighter, sometimes fainter. These *variable stars* fall into one of two broad types: *cataclysmic variables* and *intrinsic variables.* We have already discussed one type of cataclysmic variable star, the novae, which are stars that periodically change in luminosity (rate of energy output) suddenly and dramatically when they accrete material from a binary companion.

Another type of variable star is an intrinsic variable, whose variability is caused not by interaction with a binary companion but by factors internal to the star. The subset of intrinsic variable stars important in distance calculations are *pulsating variable* stars. Cepheid variable stars and RR Lyrae stars are both pulsating variable stars.

The pulsating variables vary in luminosity because of regular changes in their diameter. Why does this relationship exist? Stars are a little like ringing bells. We know when struck, large bells vibrate more slowly, producing lower tones. Tiny bells vibrate rapidly, producing higher tones. In the same way, more massive stars vibrate slowly, and smaller stars vibrate more rapidly.

A pulsating variable star is in a late evolutionary stage and has become unstable, its radius first shrinking and its surface heating. Then its radius expands and its surface cools. These changes produce measurable variations in the star's luminosity because luminosity depends on surface area, and the star is shrinking and expanding.

Astronomer's Notebook

The names of the class of variable stars, RR Lyrae and Cepheid, derive from the names of the first stars of these types to be discovered. RR Lyra is a variable star (labeled RR) in the constellation Lyra. The Cepheid class is named after Delta Cepheus, the fourth-brightest star, and a variable star in the constellation Cepheus. Cepheid variables have longer periods of brightness variation (as a class) and are more luminous than RR Lyrae stars. RR Lyrae stars have periods of less than a day. Cepheid periods range upward of 50 days.

Scientists named the two types of pulsating variables after the first known star in each group. The RR Lyrae stars all have the similar average luminosity of about 100 times that of the Sun. Cepheid luminosities range from 1,000 to 10,000 times the luminosity of the Sun. Because astronomers can recognize both pulsating variables by their pulsation pattern and average luminosity, they make convenient markers for determining distance. Cepheid variable stars are intrinsically more luminous, and Henrietta Swann Leavitt proposed that they would be useful for measuring greater distances.

Leavitt was a talented astronomer at the Harvard College Observatory who first observed a relationship between the pulsation period and the intrinsic brightness of a star. Armed with her period-luminosity relation, American astronomer Harlow Shapley (1885–1972) used the period of RR Lyrae variable stars in almost 100 globular clusters in the Galaxy to determine their distances. With this distance information, Shapley could see that the globular clusters were all centered on a region in the direction of Sagittarius, about 25,000 light-years from Earth. He concluded from his study of globular clusters that the Galaxy was perhaps 30,000 parsecs (100,000 light-years) across—far larger than anyone had ever before imagined.

Close Encounter

Henrietta Swan Leavitt was one of a number of talented astronomers on the staff of the Harvard College Observatory. She was the first to propose, in 1908, that the period of a certain type of intrinsic variable star (a Cepheid variable) was directly proportional to its luminosity.

She observed a large number of variable stars in the Magellanic clouds (companions to our own Galaxy), discovering over 1,700 of them herself. The advantage of studying these clouds is that by figuring their distance from Earth, we are able to estimate the distance from Earth to those stars within them. What Leavitt noticed was that the brightest Cepheid variable stars in the Magellanic clouds always had the longest periods, and the faintest always had the shortest periods.

What her discovery meant was huge—that astronomers could simply measure the period of a Cepheid variable star and its apparent brightness to derive its distance directly. What Leavitt accomplished was to greatly extend the astronomer's ruler from a few hundred light-years (using the parallax method) to tens of millions of light-years.

In an eerie twentieth-century replay of the days of Copernicus and Galileo, Shapley had demonstrated that the hub of the Galaxy, the Galactic center, was certainly not where we are but lay 8,000 parsecs (25,000 light-years) from the Sun. Our star and our solar system are far from the center of it all.

Each of the earlier defined parts of the Milky Way—the nuclear bulge, the disk, and the halo—has a characteristic size. Studies in the twentieth century have shown that the different parts of the Milky Way are also characterized by distinct stellar populations.

If you look at a true-color picture of the Milky Way or a similar galaxy, such as Andromeda, the bulge appears more yellow, and the disk appears more blue or white-blue. The color of a region tells us the average color of most of the stars that reside there. Globular clusters, the bulge, and the halo appear to contain mostly cooler (yellow) stars. The young (hot) stars in the globular clusters and the halo are gone, and new ones aren't taking their places in those locations.

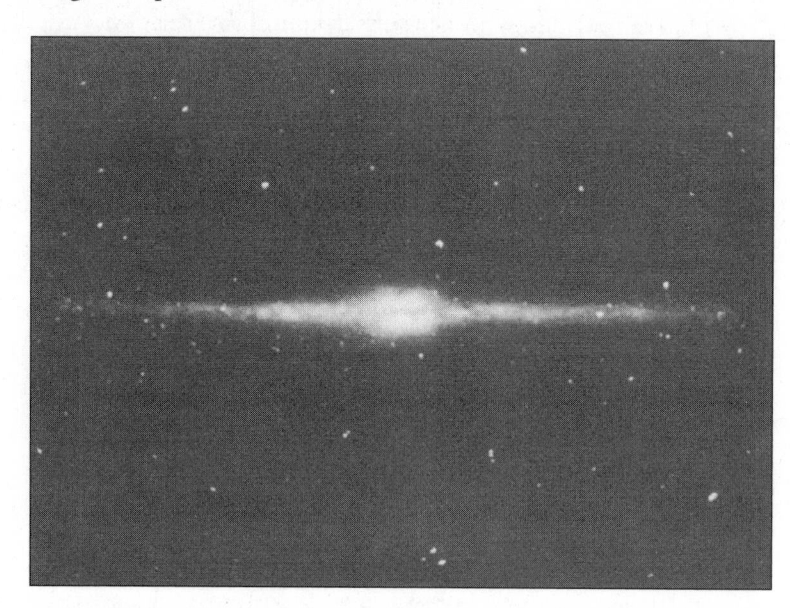

The Diffuse Infrared Background Experiment (DIRBE) on the Cosmic Background Explorer (COBE) probe made this image of the Milky Way from an edge-on perspective. The basic disk-bulge architecture of the Milky Way is apparent in this infrared image that highlights the location of low-mass stars and dust.

(Image from NASA)

The presence of large numbers of the youngest, hottest stars of type O and B makes the Galactic disk appear blue. Because O and B stars have such short lifetimes (1 to 10 million years), their presence in the disk tells us that they must be currently forming there. Remember that, at our radius, the Milky Way rotates once every 225 million years, and that means that tens of generations of massive O and B stars form and explode every time the disk of the Galaxy rotates. Cooler and smaller G, K, and M stars are also present in the disk, but the giant blue stars, far more luminous, outshine them, imparting to the entire disk region its characteristic blue-white color.

Why are the youngest stars in the disk and the oldest in the halo?

The Galactic disk is where the interstellar gas clouds reside, the so-called giant molecular clouds (GMCs). With raw materials plentiful, star production is very active here; therefore, young stars are abundant. In contrast, the halo region has very little nonstellar material, so no new stars are being created there. In the Galactic bulge, there is an abundance of interstellar matter as well as old *and* new stars.

Milky Way Portrait

The different parts of the Milky Way are not static but are in constant motion. The disk rotates about the Galactic center, and at large radii, the rate of rotation does not trail off but remains fairly constant. This rotation is in contrast to what we see in a planetary system, where planets rotate more and more slowly the farther they are from the center. The constant rotation rate at large distances from the Galactic center betrays the presence of material we cannot detect with normal observations.

The stars in the halo move very differently, plunging through the Galactic disk in randomly oriented elliptical orbits, which are not confined to the Galactic disk and in some ways seem unaware of it. All of these orbits are centered on what we call the Galactic center region and give clues as to how the Galaxy formed.

Astro Byte

Our part of the Galactic disk orbits the Galactic center at over 136 miles per second (220 km/s), taking about 225 million years for our region to complete one orbit around the Galactic center. Our solar system has orbited the Galactic center some 15 to 20 times since it formed. One quarter of a Galactic orbit ago, dinosaurs roamed Earth.

The Birth of the Milky Way

The structure, composition, and motion of the Milky Way hold the keys to its origin. Though the theory of galaxy formation is far from complete, we have a fairly good picture of how our Galaxy might have formed.

Ten to fifteen billion years ago, an enormous cloud of gas began to collapse under its own gravity. The cloud, like most of the universe, would have been mainly hydrogen. It would have had a mass equal to that of all the stars and gas in the Milky Way—several hundred billion solar masses.

The stars that formed first as the great cloud collapsed assumed randomly oriented elliptical orbits with no preferred plane. Today, the oldest stars in our Galaxy, which are those in the Galactic halo and Galactic bulge, ring with the echoes of the early days. They are the "asteroids" of our Galaxy, and the Galactic halo is a vestige, a souvenir of the Galaxy's birth. Indeed, the globular clusters in the halo may even have formed prior to the cloud's collapse.

In a gravitational collapse process very similar to that which created the solar system, the great cloud began rotating faster around a growing mass at the center of the Galaxy. The rotation and collapse caused the clouds of gas to flatten into the Galactic disk, leaving only those early stars in the halo.

Because the Galactic disk is the repository of raw materials, it is the region of new star formation in the Galaxy. The Galactic halo is out of fuel and consists only of older (cool) redder stars. The orderly rotation of stars and gas in the Galactic disk stands in contrast to the randomly oriented orbits of stars in the halo and the Galactic bulge.

Dark Matters

"What you see is what you get," the popular saying goes. However, such is not always the case in astronomical affairs.

Using Kepler's Third Law, we can calculate the mass of the Galaxy. This mass (expressed in solar masses) can be derived by dividing the cube of the orbit size (expressed in astronomical units, or A.U.) by the square of the orbital period (expressed in years). And Isaac Newton told us that at a given radius, all of the mass causing the rotation can be considered to be concentrated at a point at the center of rotation. For a system in "Keplerian rotation" (like a planetary system), we would expect the velocities of rotation to decrease as we looked farther and farther out. Jupiter, for instance, orbits more slowly than Mercury. Taking this approach, we find that the mass of the Milky Way *within* 15,000 parsecs of the Galactic center—that is, the *radius* of the visible Galaxy (diameter ~30,000 parsecs)—is 2×10^{11} solar masses or about 200 billion solar mass stars.

Common sense suggests that the mass of the Galaxy drops off precipitously as we run out of matter moving toward the visible outer edge of the Galaxy. Yet the puzzling fact is that *more* mass is contained *beyond* this boundary than within it. So much for common sense! What could be going on?

Within a radius of 40,000 parsecs from the Galactic center, the mass of the Milky Way is calculated to be about 6×10^{11} solar masses: 600 billion solar masses. This means that as much, if not more, of the Milky Way is *unseen* as is *seen.* Spiral galaxies, such as the Milky Way, have mass-to-light ratios ranging up to 10, even in the visible part of their disks. That is, luminous matter on very large scales accounts for only 10 percent of the matter that we can "see" via gravity.

So what is all this other stuff we cannot see directly? Whatever its makeup, it apparently emits no radiation of any kind—no visible light, no x-rays, no gamma radiation. But it cannot hide completely. We "see" it because it has mass and its mass affects the way in which the stars and gas of the Milky Way orbit.

Astronomers call the region containing this mass the "dark halo," and the Milky Way is not alone in possessing such a region. Many, if not all, galaxies have the same signature in the rotation of their stars and gas. Presumably, the dark halo contains—what else?—*dark matter*, a catchall term that we use to describe a variety of candidate objects. The truth is, we're not sure what dark matter is, but we do know it's there because we clearly can see its gravitational effects.

The nature of dark matter is one of the greatest mysteries of science. Some astronomers have suggested that difficult-to-detect low-mass stars (brown dwarfs or faint red dwarfs) might be responsible for the mass in this region—although recent Hubble Space Telescope observations have suggested an insufficient quantity of such objects to account for so much mass. It has recently been established that neutrinos do have nonzero mass, so their presence might contribute to dark matter. Yet others propose that dark matter consists of hitherto unknown subatomic particles, which pervade the universe. Nature has yet to reveal this particular mystery.

In the Arms of the Galaxy

We have referred to our own Galaxy as a spiral galaxy, but what are the spiral arms that, observed from afar, appear to be arcs of bright emission, curving out from the center of the Galaxy? They are among the most beautiful sights in the universe. Some spiral galaxies have two arms, others more, and some arms are loosely wound, others tightly.

How do we know that the Milky Way consists of spiral arms? Using the 21 cm hydrogen line, astronomers have plotted the distribution of neutral hydrogen (by far the most abundant element in the Galactic disk or anywhere else). Both position and velocity of the hydrogen clouds are required to make this plot because it shows a dimension that is not on the sky, namely depth. These radio images confirm the spiral structure of the Milky Way.

Radio studies also confirm that we are far from the Galactic center, located unostentatiously on a cusp between two spiral arms. Spiral arms themselves are not that hard to account for, but their longevity is. In the next chapter, we discuss how spiral arms might arise, and how they could possibly persist.

A detail of the plane of our Galaxy as seen in neutral hydrogen (HI) comes from one of the projects mapping our Galaxy, the Canadian Galactic Plane Survey (CGPS). This image is of a star-forming region in the direction of the constellation Cygnus.

(Image from J. English, A. R. Taylor/CGPS)

Is There a Monster in the Closet?

The central part of our Galaxy (first identified by Harlow Shapley) is quite literally invisible. We cannot observe it at optical ("visible") frequencies, but radio and infrared frequency observations have told us much about this hidden realm.

Recent radio-frequency observations have identified a ring of molecular material orbiting the Galactic center at a distance of perhaps 8 or so parsecs: 25 light-years. The rotational velocity of this ring tells us that there are several million solar masses of material located within its radius. Other radio observations (sensitive to hot gas) show that the ring itself contains only a small amount of ionized material, nowhere near a few million solar masses.

But we have more clues: radio astronomers have picked up strong emissions from a source in the Galactic center region called Sagittarius A* (pronounced "A-star") that appears to be tiny—only 1 A.U. across, the size of the orbit of Earth around the Sun!

This source appears to be the location of a black hole of several million solar masses and is the same source that has been imaged with the new orbiting Chandra X-ray Observatory. The x-ray emissions from the Galactic center show the presence of hot gas and a jet, both of which are often associated with the presence of a black hole.

Finally, recent infrared observations show that the Galactic center region contains a great many stars, closely packed together. The stars cannot account for the required mass, but their high-velocity orbits tell us something else about the tiny Sagittarius A* source. It contains all that mass!

This x-ray image of the Galactic center taken with the Chandra X-ray Observatory shows the presence of two lobes of hot gas (labeled) perpendicular to the Galactic plane. These types of features are often associated with the presence of a black hole.

(Image from Chandra/NASA)

Careful tracking of the orbits of stars in the infrared around Sagittarius A* have shown that there must be a very massive, very compact object located there. Stellar orbits tell us that the mass the stars are orbiting is 2.6×10^6 times the mass of the sun. These high velocities, and the tiny size of Sagittarius A*, suggests that the stars are orbiting a supermassive black hole: a monster in our closet.

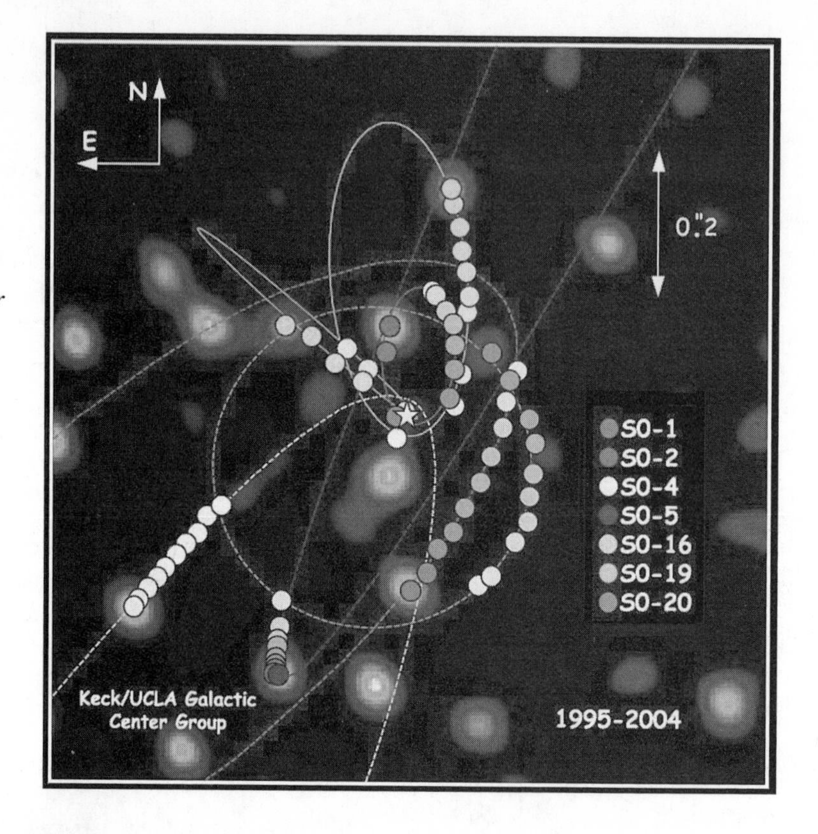

This image shows the orbits of stars near the center of our own Galaxy, the Milky Way. The sizes of the orbits indicate that the stars are orbiting something very massive and very compact. These orbits are some of the strongest evidence to date for a black hole.

The Least You Need to Know

- Our home Galaxy is the Milky Way, which we can see from within as an arc of fuzzy emission across the sky.
- The main components of the Galaxy are the disk, the stellar halo (containing the globular clusters), the nuclear bulge, and the dark halo.
- Our solar system is in the Galactic disk of the Milky Way, 25,000 light-years from the Galactic center.
- We can only account for a fraction of our Galaxy's mass with normal matter; most of the Galaxy's mass must be contained in a halo of material, consisting of "dark matter."
- Astronomers have built a map of the gas in the disk of our Galaxy using radio telescopes.
- A large amount of evidence suggests that the center of our Galaxy harbors a massive black hole.

Chapter 14

A Galaxy of Galaxies

In This Chapter

- Hubble's classification of galaxy types: spiral, elliptical, and irregular
- Spiral density waves
- Determining the distance of galaxies
- Galactic clusters and superclusters
- Calculating the mass of galaxies and galactic clusters
- The nature of Gamma Ray Bursts (GRBs)
- Hubble's Law

Viewed with the naked eye, some "stars" appear fuzzier than others. After the invention of the telescope, we understood why this was so. Through a telescope, some apparent "stars" are resolved into the disks of planets, some into regions where stars are forming, and others into collections of old stars. One class of objects identified in the eighteenth and nineteenth centuries, called *spiral nebulae*, caused great disagreement among astronomers. Clearly, these were not stars, and yet there was no way to figure out how big they were without knowing how far away they were.

This type of disagreement has occurred again and again in astronomy. More recently, the nature of Gamma Ray Bursts (GRBs) has been debated, but the issue is the same. Are GRBs lower-energy, relatively nearby phenomena, or are they incredibly energetic, more distant phenomena? Later in this chapter, we explain they are, in fact, the final moments of stars in distant galaxies.

Before Edwin Hubble extended our conception of the size of the universe in the 1920s, we classified all these fuzzy objects as nebulae—because they were, well, nebulous—and they were generally thought to lie within the Milky Way, which, in turn, we believed to be synonymous with the entire universe. We now know that our Galaxy is but one of many and that the spiral nebulae are entire other galaxies, some smaller and some bigger than our own, all containing hundreds of billions of stars. And we see galaxies no matter where we look in the sky. When we look deep enough, they are there. Some are so close they are part of our "Local Group"; others are so distant their light has had barely enough time to reach us since the universe began.

In this chapter, we talk about galaxies other than our own. Astronomers have two major goals in the exploration of other galaxies. The first is to understand them and how they have evolved with time. The second is to use them as models to tell us more about the Milky Way, our own galactic home.

Sorting Out the Galaxies

Each galaxy contains several hundred billion stars, typically about 100 times as many stars as there are people on our planet.

And many galaxies are out there.

Fortunately, despite their mind-numbing numbers, galaxies are not all completely different from one another, but they fall into broad groups that we can classify, just like trees or beetles or clams. But they couldn't be classified until we photographed large numbers of them, starting in the mid- to late nineteenth century, after the advent and development of astronomically useful photographic techniques.

The great American astronomer Edwin Hubble (1889–1953) came to the Mount Wilson Observatory in California in 1922 to use its new 100-inch (2.5-meter) telescope to study nebulae. A fundamental question remained unanswered back then. Were these spiral nebulae as big as our Galaxy and incredibly distant, as first suggested by Immanuel Kant in his book *Universal Natural History and Theory of the Heavens* in the eighteenth century? Or were they more mundane, relatively nearby objects? The huge surface area and high resolution of the 100-inch telescope allowed Hubble to see that these nebulae, like our own Milky Way, resolved into individual, faint stars.

After intrinsic variable stars were identified in some of these "nebulae," astronomers could use the relationship between period and luminosity to determine their distance. Calculating distances from Cepheid variable stars, Hubble stunningly concluded in 1924 that many of the nebulae were not part of the Milky Way at all but were galaxies in their own right. He thought Andromeda, for example, was almost 1 million light-years away, *far* outside the limits of our own Galaxy; this distance has since been revised upward to 2.5 million light-years. Shortly after making his discovery, Hubble, a man not easily overwhelmed, set about classifying the galaxies he saw.

Like many first impressions, Hubble based his classifications on appearance. He established three broad categories: spiral, elliptical, and irregular. It wasn't until later that the explanation for the different classes was understood.

Spirals: Catch a Density Wave

In the preceding chapter, we took a long look at one example of a spiral galaxy, the Milky Way. Hubble labeled all spiral galaxies with the letter "S" and added an a, b, or c, depending on the size of the galactic bulge. Sa galaxies have large bulges, Sb medium-sized bulges, and Sc the smallest bulges. Very clearly defined, tightly wrapped spiral arms are also associated with Sa galaxies, whereas Sb galaxies exhibit more diffuse arms, and Sc galaxies have even more loosely wrapped, less clearly defined spiral arms.

A region in NGC 1365, this barred-spiral galaxy is located in a cluster of galaxies called Fornax. The central image was photographed through a ground-based telescope. The images to the left and right show the central bulge as seen at optical and infrared wavelengths with the Hubble Space Telescope.

(Image from NASA and M. Carollo)

Another spiral subtype is the barred-spiral galaxy, which exhibits a linear bar of stars running through the galactic disk and which bulge out to some radius. In barred-spiral galaxies, the spiral arm structure typically begins at the ends of the bar. The barred-spiral subgroup, designated SB, also includes the a, b, and c classifications, based on the size of the galactic bulge and the winding of the arms. The Milky Way Galaxy appears to have a bar structure near its center.

Ellipticals: Stellar Footballs

Elliptical galaxies present a strikingly different appearance from the spirals. When viewed through a telescope, they look a bit dull compared to their flashy spiral cousins. They have no spiral arms nor any discernable bulge or disk structure. Typically, these galaxies appear as nothing more than round or football-shaped collections of stars, with the most intense light concentrated toward the center and becoming fainter and wispier toward the edges.

Of course, the orientation of an elliptical galaxy influences the shape we see in the sky; that is, the apparent shape of a given galaxy might not be its true (intrinsic) shape. Consider a football. Viewed from the side, it has a sort of oval shape, but when viewed end-on, it looks like a circle. Regardless of true shapes, Hubble differentiated within this classification by using apparent shape. E0 ("E-zero") galaxies are almost circular, and E7 galaxies are very elongated, or elliptical. The rest—E1 through E6—range between these two extremes. Astronomers can use observations of the motions of stars in elliptical galaxies, combined with computer modeling, to determine the true shape of a particular elliptical galaxy.

An elliptical galaxy has no such thing as a typical size. Their diameters range from a thousand parsecs (these dwarf ellipticals are much smaller than the Milky Way) to a few hundred thousand parsecs (giant ellipticals). The giant ellipticals, often located in the center of galaxy clusters, are many times larger than our Galaxy.

Like any classification scheme, Hubble's has its share of duck-billed platypuses, objects that don't quite fit in. Some elliptical-shaped galaxies exhibit more structure than others, showing evidence of a disk and a galactic bulge. They still lack spiral arms, and like the other ellipticals, they don't contain star-forming gas. This type of galaxy (since it does have a disk and a bulge) is designated S0 (pronounced "S-zero"). Some of these galaxies even contain a bar and are designated SB0.

Hickson Compact Group (HCG) 87 consists of four galaxies. The edge-on spiral at the bottom of the image and the elliptical to the right are both known to have "active nuclei," most likely related to the presence of a black hole.

(Image from AURA/STScI/NASA)

Are These on Sale? They're Marked "Irregular"

Finally comes the miscellaneous category of irregular galaxies, which lack any regular structure. Galaxies in this class typically look as if they are coming apart at the seams or are just plain messy. Unlike the ellipticals, irregulars are rich in interstellar material and are often sites of active star formation.

Close Encounter

Those of us confined to Earth's northern hemisphere don't get to see the most spectacular of the irregular galaxies, the Small and Large Magellanic Clouds. Named for the sixteenth-century explorer Ferdinand Magellan, whose men brought word of them to Europe at the conclusion of their global voyage, the Magellanic Clouds are rich in hydrogen gas. The clouds, like moons, are believed to orbit the much larger Milky Way. Like any object with mass in the universe, galaxies feel the irresistible tug of gravity and are pulled into groupings by its effects.

Two of the most famous irregular galaxies are very close to us: the Large and Small Magellanic Clouds. Usually called the LMC and the SMC by acronym-happy astronomers, they are visible from the southern hemisphere. These Milky Way companions (10 times closer to us than the closest spiral galaxy) interact with each other and our Galaxy via gravity, and both contain active star-formation regions. The Small Magellanic Cloud is stretched into an elongated shape due to the tidal forces exerted on it by our Galaxy.

Galactic Embrace

In the constellation Canes Venatici is Messier object 51 (M51), the Whirlpool galaxy. Its distinct shape resembles an overhead view of a hurricane or, less dramatically, a pinwheel. Bright curved arms grow out of its hublike galactic bulge and wrap partway around the galactic disk. It is easy to see the spiral pattern in the Whirlpool galaxy because it faces us at an angle that gives us a spectacular "overhead" view rather than a side view, but what exactly are those characteristic spiral arms?

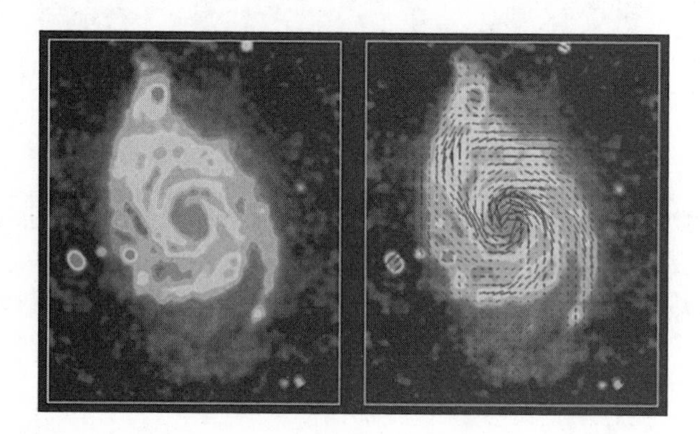

The spiral Whirlpool galaxy M51 is seen with the Very Large Array (VLA) and the Effelsberg 100-m at 6 cm. The short lines in the image to the right show the orientations of the magnetic fields embedded in material in the spiral arms.

(Image from NRAO/AUI)

In fact, the very existence of the arms does present a puzzle. We know the stars and other matter in the galactic disk orbit differentially—faster toward the center, somewhat slower toward the periphery. This rotational pattern soon stretches any large clump of stars or gas in the disk of a spiral galaxy into a spiral structure. It does this so quickly that if differential rotation were to account for spiral arms, the arms would rapidly get wrapped up around a galaxy's bulge and disappear. Somehow, the Milky Way and other spiral galaxies retain their spiral arm structure for long periods of time, long enough that they are plentiful in the universe.

Most astronomers are now convinced that the arms of spiral galaxies are the result of compressions, called *spiral density waves*, moving through a galaxy's disk. These ripples in a galaxy's disk move around the galactic center, compressing clouds of gas until they collapse and form stars. The ionized gas surrounding hot, young stars, formed by the passage of a density wave, is what we see as spiral arms in the optical part of the spectrum.

The theory of spiral density waves neatly resolves the problem of the effect of differential rotation. A wave is a disturbance that moves through matter, like a ripple in a pond. Thus, spiral density waves move *through* the matter of a galactic disk and are not caught up in differential rotation. The stars they form might get stretched out by this differential rotation, but the waves keep on moving.

A close encounter between two galaxies is one way to trigger a density wave. Computer simulations clearly show featureless disks developing spiral structures as two galaxies approach one another. Galaxy collisions are slow-motion train wrecks, taking hundreds of millions of years to occur, but we see evidence of their occurrence littering the universe.

How to "Weigh" a Galaxy

In the same way we cannot directly measure the temperature of stars, we cannot directly measure their mass. We measure the mass of the Sun, for instance, by using Kepler's Third Law. If we know the distance to the Sun and the period of Earth's orbit, we can calculate the Sun's mass. We can calculate galaxy masses in the same way, only the objects orbiting are stars and gas instead of planets.

A Big Job

Plotting the *rotation curve* of an individual spiral galaxy—the velocity of disk material versus the distance of that material from the galactic center—can yield the mass of the galaxy that lies within that radius. This method gives a good estimate as long as most of a galaxy's mass is contained near its center. In the solar system, for example, the outer planets rotate much more slowly than the inner planets, in accordance with Kepler's Third Law because the mass of the solar system (99.9 percent of it, anyway) is contained in the Sun.

But astronomers soon noticed a problem with galaxies. Objects in the outer reaches of spiral galaxies (clouds of gas—in particular, clouds of neutral hydrogen called HI clouds) orbit in a way that indicates they "see" more mass out there than we do here at the radius of the Sun. That is, they orbit faster than they should. Using the 21 cm radio line of hydrogen to see the HI, astronomers have traced the rotation curve of many galaxies far beyond the outermost stars. These curves seem to indicate that there is more matter at large radii in these galaxies. Whatever has this mass, though, is something we can't see because it is not "shining" at any wavelength. The rotation of galaxies indicates the presence of that mysterious stuff astronomers dub dark matter.

Astronomer's Notebook

Most spiral galaxies contain from 10^{11} to 10^{12} solar masses of matter, much of it located far beyond the visible radius of the galaxy. Large elliptical galaxies contain about the same mass (but can be even more massive), whereas dwarf ellipticals and irregular galaxies typically contain between 10^6 and 10^7 solar masses of material.

On the scale of clusters of galaxies, the mass-to-light ratio, which is about 10 in galaxies, can be 100 or more. That is, luminous matter on very large scales accounts for only 1 percent of the matter that we "see" via gravity.

"It's Dark Out Here"

Because dark matter accounts for a great deal of the mass of a galaxy—up to 10 times more than the mass of visible matter—the shocking conclusion must be that 90 percent of the universe is dark matter, utterly invisible in the most profound sense of the word. Dark matter neither produces nor reflects any electromagnetic radiation of any sort at any wavelength.

You can think of it as an embarrassing truth or as an exciting unanswered question. The fact remains: we're not quite sure what 90 percent of the matter in the universe is made of.

More Evidence, Please

One galactic collision in a region of space called the Bullet Cluster (galaxy cluster 1E0657-56) has provided additional evidence for the presence of dark matter. In this galactic train wreck, observed with both the Hubble Space Telescope and the Chandra X-ray Observatory, the stars and dark matter have torn past one another, and the gases have mixed together. The result is a patch of mixed gases in the middle and two patches of stars on either side.

Mixed in with the stars is more mass than the stars themselves contain, and because the gas has been stripped away by the collision, the only explanation for the mass is dark matter. These observations still don't tell us what dark matter is; they just provide more evidence that it is there and that it is mixed in with the stars.

Let's Get Organized

Human beings are inveterate pattern makers. Men and women have looked at the sky for centuries and have superimposed upon the stars patterns from their own imaginations. Although the mythological heritage of constellations is still with us, scientists long ago realized that the true connections among the planets and among the stars are matters of mass and gravitational force, not likenesses of mythological beings.

So are there gravitationally determined patterns to the distribution of galaxies throughout the universe?

Measuring Very Great Distances

By radar ranging, we can accurately measure the distance from Earth to the planets, but measuring the distance to farther objects, namely the nearer stars, requires measuring stellar parallax. However, beyond about 100 parsecs, parallax doesn't work well because the apparent angular displacement becomes smaller than the angular resolution of our best telescopes.

We can use gas velocities and a model of the rotation of the Milky Way to measure distances within our own Galaxy out to about 30,000 light-years. Beyond this, and out to 10 to 20 million parsecs (30 to 60 million light-years), we can identify and observe variable stars (the RR Lyrae and Cepheid variables) in order to determine distance.

The trouble is that Cepheid variable stars farther than 15 million parsecs are difficult to resolve and, for most telescopes, too faint to be detected. And many, many galaxies are well beyond 15 million parsecs away.

To get around this, astronomers have used two methods to estimate intergalactic distances greater than 15 million parsecs. One tool is called the Tully-Fisher relation, which uses an observed relationship between the rotational velocity of a spiral galaxy and its luminosity. By measuring how fast a galaxy rotates (astronomers use the 21-cm hydrogen line), one can calculate its luminosity with remarkable accuracy. Once we know the luminosity of a galaxy, we can measure its apparent brightness and easily calculate its distance out to several hundred million parsecs.

The other tool is more general and involves identifying various objects whose luminosity we know. Such objects are referred to as *standard candles.* If we truly know the brightness of a source, we can measure its apparent brightness and determine how distant it is. A 100-watt light bulb is an example of a standard candle. It will look fainter and fainter as it recedes from us, getting dimmer as distance squared, and in fact, we can determine how far it is by measuring its apparent brightness versus its known 100-watt luminosity.

def•i•ni•tion

A **standard candle** is any object whose luminosity is well-known. We can then use its measured brightness to determine how far away the object is. The brightest standard candles can be seen from the greatest distances.

Gamma Ray Bursts as Candles

Gamma Ray Bursts (GRBs) have been observed for decades, but only recently have astronomers come to a consensus as to what they might be. Observationally, they are point sources of gamma ray emission that appear to be located at very great distances. So if we can understand the physics of GRBs, we have a hope of extending our "distance ladder" very far indeed—out farther than the highest redshift objects yet observed. They can be a million to a billion times brighter than supernovae.

The bursts of gamma rays are now thought to be associated with the death of a massive star and the subsequent birth of a black hole. GRBs should allow us to see farther back in time, closer to the time of the Big Bang than ever before.

One very interesting standard candle is a Type Ia supernova. It is "interesting" because its peak luminosity is very regular, not to mention enormous: 10 billion or 10^{10} solar luminosities. We have discussed the Type II supernovae, the core collapse of a massive star. Type Ia supernovae occur when a white dwarf accumulates enough material from its binary red giant companion to exceed 1.4 solar masses. When this happens, the white dwarf begins to collapse, and the core ignites in a violent burst of fusion. The energies are sufficient to blow the star apart, producing an event so luminous we can see it *billions* of light-years away.

The Local Group and Other Galaxy Clusters

Armed now with the ability to measure very great distances, we can begin looking at the relationships among galaxies.

Within 1 million parsecs (3 million light-years) of the Milky Way lie about 20 galaxies, the most prominent of which is Andromeda (M31). This galactic grouping, called the Local Group, is bound together by gravitational forces.

The generic name for our Local Group is a galaxy cluster, of which thousands have been identified. Some clusters contain fewer than the 20 or so galaxies of the Local Group, and some contain many more. The Virgo Cluster, an example of a rich cluster, is about 15 million parsecs from the Milky Way and contains thousands of galaxies, all bound by their mutual gravitational attraction. Giant elliptical galaxies are often found at the centers of rich clusters where there are very few spirals.

The galaxy cluster called Abell 2218 is about 2 billion light-years from Earth. When these photons left the cluster, life on Earth was very simple indeed. The effect of the "unseen" mass in this cluster can be seen as the arcs of light that are distorted light from background sources. This distortion, much like viewing a room through a wine glass, is sometimes referred to as "lensing."

(Image from NASA, A. Fruchter, and the ERO team)

From the velocities and positions of galaxies in clusters, one thing is very clear. We cannot directly observe at least 90 (perhaps as much as 99) percent of the mass that must be there. Galaxy clusters, like the outer reaches of spiral galaxies, contain mostly dark matter.

Superclusters

Galaxy clusters themselves are grouped together into what we call superclusters. The Local Supercluster, which includes the Local Group, the Virgo Cluster, and other galaxy clusters, encompasses some 100 million parsecs.

On the very largest scales we can measure (hundreds of millions of light-years across), the universe has an almost bubbled or spongy appearance, with superclusters concentrated on the edges of large empty regions or voids. In the final chapters of this book, we explore the possible explanations for this large-scale structure.

Where Does It All Go?

The galaxies within a cluster don't move in an orderly manner. Like the stars in the galactic halo that envelops a spiral galaxy, the galaxies appear to orbit the cluster center in randomly oriented trajectories. But on larger scales, beyond the confines of a single cluster, there does appear to be orderly motion, which is the echo of a cataclysmic event: the event that brought the universe into being.

Hubble's Law and Hubble's Constant

We have known since the early twentieth century that every (sufficiently distant) spiral galaxy observed exhibits a redshifted spectrum, indicating that all of these galaxies are moving away from us. Recall the Doppler effect: wavelengths grow longer (they redshift) as an object recedes from the viewer. The conclusion is inescapable: all galaxies partake in a *universal recession*. This is the apparent general movement of all galaxies away from us. This observation doesn't mean we are at the center of the expansion. Any observer located anywhere in the universe sees the same redshift, perceiving that the galaxies are all moving away from his or her point of view.

In 1929, Edwin Hubble and Milton Humason first plotted the distance of a given galaxy against the velocity at which it receded. The resulting plot was dramatic. The rate at which a galaxy is observed to recede is directly proportional to its distance from us; that is, the farther away a galaxy is from us, the faster it travels away from us. This relationship is called *Hubble's Law*. It relates the velocity of galactic recession to its distance from us. Simply stated, Hubble's Law says that the recessional velocity is directly proportional to its distance from the observer.

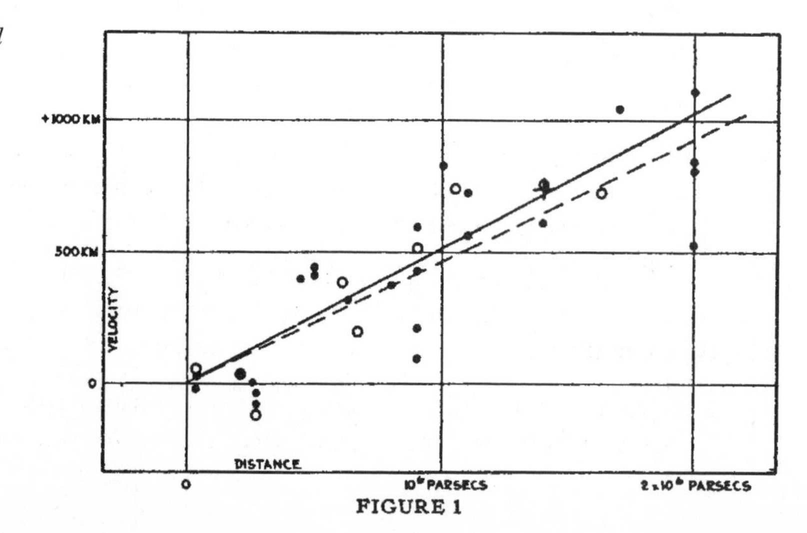

This plot is Hubble's original figure from his 1929 paper, showing the radial velocities in km/sec on the x-axis and his estimates of their distances (in parsecs) on the y-axis.

(Image credit E. Hubble)

To imagine what this expansion describes, picture a bunch of dots on the surface of a toy balloon. Our Galaxy is one of the dots, and all of the other galaxies in the universe are the other dots. As you inflate the balloon, the surface of the balloon stretches, and, from the point of view of any dot (galaxy), all the other dots (galaxies) are moving away. The farther away the dot, the more balloon there is to stretch, so the faster the dot will appear to recede.

One benefit of Hubble's Law is that we can use it to extend our cosmic distance scale to extraordinary distances. Although the Tully-Fisher technique will get us out to about 200 million parsecs, and Type Ia supernovae to a few billion parsecs, Hubble's Law goes farther. In fact, Hubble's Law gives us the distance to any galaxy for which we can measure a spectrum (in order to get the Doppler shift). With the velocity from that spectrum and the correct value of *Hubble's constant*, we arrive quite simply at the distance.

But there is a twist. We have assumed that the universal expansion happens at a constant rate. Is this a good assumption?

Recent observations suggest that it is not. As far as we can now tell, the universe has not always been expanding at the same rate. Two magnificent forces are in opposition. There is the expansion of the universe (as we will see, set into play by the Big Bang), and there is the force of gravity, pulling every two particles with mass together, diminished by distance squared. If nothing was pushing things apart, the universe would, due to gravity, collapse to a point. This would be a problem, and Einstein recognized it as such. When he was working on general relativity, he needed to add a term—he called it the "cosmological constant"—to his equations in order to balance the force of gravity, to keep the universe "puffed up." This term was needed to keep the universe static, as Einstein and everyone else then thought it to be.

def•i•ni•tion

Hubble's Law turns on **Hubble's constant** (H_0), the constant of proportionality between the velocity of recession and the distance from us. The value of H_0 is expressed in kilometers per second per megaparsec (a megaparsec [Mpc] is 1 million parsecs) and is found to be about 65 km/s/Mpc, meaning that the universe has an age of about 14 billion years.

Einstein's Blunder

But then Edwin Hubble came along and showed that the universe was hardly static; in fact, everything was rushing away from everything else, and although gravity might eventually win out, the universe wasn't going to collapse to a point anytime soon. Einstein later called the introduction of the cosmological constant into his equations his "greatest blunder."

As it turns out, however, Einstein's "blunder" might not have been a mistake at all. The cosmological constant might actually be needed. Recent observations made by scientists of very distant ("high-redshift," in astronomer parlance) Type Ia supernovae show something very surprising. These "standard candles" can be seen out to 7 billion light-years, and the distances to the galaxies that they are in show that the universe appears to have been expanding more slowly in the past than now. So not only is everything rushing away from everything else, but, these days, it's also rushing away more quickly. In other words, the expansion of the universe is accelerating.

And what is causing it to accelerate? Well, it might be the "vacuum energy" that Einstein thought he needed to support the universe from collapsing on itself. Interestingly, this force increases with distance—as opposed to gravity, which weakens with distance. Scientists have dubbed this new energy required to power an accelerating expansion "dark energy," implying that, as with "dark matter," we know it is there, but we don't know what it is.

The Big Picture

Expansion implies a beginning in time. In the coming chapters, we explore where and how the expansion might have begun and the details of the expansion itself. But before we move on to some of these big questions, let's turn our attention to some of the most energetic and unusual members of the galactic family: the quasars, black holes, and galactic jets, all of which keep the universe quite active, thank you very much.

The Least You Need to Know

- Edwin Hubble first classified galaxies into three broad types: spiral, elliptical, and irregular.
- The majority of the mass (90 percent) of most galaxies and clusters (99 percent) is made of dark matter, material we cannot observe but only see by its gravitational effect.
- Although several lines of evidence exist that confirm dark matter is very real, we are still unsure of what it might be.

- Many galaxies are grouped in galactic clusters, which, in turn, are grouped into superclusters that are found together on the edges of huge voids.
- Edwin Hubble first observed that galaxies recede from us at a rate proportional to their distance from Earth; the constant of proportionality, called Hubble's constant, tells us the age of the universe.
- The expansion of the universe appears to be accelerating, powered by an as yet unexplained "dark energy."

Part 5

The Big Questions

The questions themselves may be expressed simply enough: Is the universe infinite or finite? Eternal or tied to time? If it had a beginning, when and how did it begin? And how will it end? Or will it end at all? Finally, in this universe, are we alone? Is there life—other than on Earth—within the solar system? Within the Galaxy? Beyond? If so, is it intelligent? And if intelligent, would it communicate with us?

The questions may be put simply enough. The answers … well, read on.

Chapter 15

Strange Galaxies

In This Chapter

- Distant galaxies, strange galaxies
- Active galaxies: Seyfert and radio galaxies
- The driving force of an active galaxy
- Galactic jets
- Composition of a quasar

During the Depression, Grote Reber, an out-of-work radio technician, decided to kill time by building his own radio telescope in his Wheaton, Illinois, backyard. He assembled his instrument in 1936, and by the 1940s, Reber had discovered the three brightest radio sources in the sky. He didn't know it at the time, but two of them—the Galactic center, Sagittarius A; and the shrapnel of a supernova explosion, Cassiopeia A—are sources in our own Galaxy. However, the third radio source, called Cygnus A, turned out to be much, much farther away and far stranger.

In 1951, the pioneering German-American radio astronomers Walter Baade and Rudolph Minkowski located a dim optical source at the position of Cygnus A and, from its spectrum, measured a redshift (or its recessional velocity) of some 12,400 miles per second (almost 20,000 km/s). Whatever it was, Cygnus A was moving away from us—fast! For a while, astronomers tried to figure out how an object in our Galaxy could be moving so swiftly. It turned out Cygnus A wasn't in our Galaxy at all but very far away. Later, astronomers discovered that this was something never before seen: a distant inferno churning out the energy of 100 normal galaxies.

In this chapter, we examine objects like Cygnus A and the other behemoths that hide deep in the hearts of distant galaxies. They are some of the most energetic and bizarre objects in the universe.

A Long Time Ago in a Galaxy Far, Far Away ...

Because of their great distances—hundreds of millions of parsecs away—the farthest "normal" galaxies are very faint. At such extreme distances, it becomes difficult even to see such galaxies, much less study their shapes or how they are distributed in space. But as far as we *can* see, little difference exists between normal distant galaxies and normal galaxies closer to home.

The operative word here is *normal.* Out in the farthest reaches of space, we see some objects that are not normal—at least they are not what we're accustomed to seeing in our cosmic neighborhood.

What does this tell us? Remember, the universe has a speed limit, the speed of light, and information can travel no faster than this speed. Thus the farther away a certain star or galaxy might be, the longer it takes its light, and the information it contains, to reach us. Thus not only are we seeing far into space, we are seeing *back into time.* The strange objects that lurk in the distant universe existed in its earliest times. To study these objects—quasars—is to peer into the origin of galaxies, including perhaps the origin of our own Milky Way.

Quasars: Looks Can Be Deceiving

After seeing the detailed, high-resolution optical images of emission nebulae and galaxies such as Andromeda, a quasar can make a disappointing first impression. Although we can see the optical counterparts of the brightest of them with an amateur telescope on a dark night, they are undistinguished sources of visible light. In fact, they look so much like stars that astronomers at first thought they were simply peculiar stars, and so dubbed them "quasi-stellar objects."

But keep in mind how incredibly distant these objects are. The closest quasars are some 700 million light-years away. Though their apparent brightness might be small, their luminosities—the amounts of energy that they put out each second—are astounding.

Quasars appear optically faint for two reasons: they emit much of their energy into the nonvisible part of the spectrum, and they are very distant objects.

The great distance to quasars became truly apparent in the 1960s. The first quasars were discovered at radio frequencies, and optical searches at these locations showed objects that looked like stars. But the spectra of these "stars" told a different story. In the early 1960s, the astronomer Maarten Schmidt made a stunning proposal. The four bright spectral lines that distinguish hydrogen from the rest of the elements in the universe were seen to be shifting to longer wavelengths—*much* longer wavelengths. These redshifted lines, along with Hubble's Law, indicated that quasars were very distant—billions of light-years away, in fact.

Close Encounter

In order to see objects that are very distant, such as quasars, astronomers need large-diameter telescopes, which collect more light and have more resolution than smaller ones. A project under way, called OWL for the OverWhelmingly Large Telescope, is now garnering funding for construction in the coming decade. This telescope will be truly staggering, having a diameter from 60 to 100 meters, which presents a formidable technical challenge, since this optical telescope will be as large as the largest radio telescope. Just consider that optical wavelengths are tiny compared to radio wavelengths, and the surfaces of the mirror must be perfect. The plan is to build the mirror in 3,048 segments, each 1.6 m across—almost as big as the Hubble Space Telescope mirror! In the plan, the secondary mirror will be over twice the size of the Keck telescope, and the "active" mirror, which corrects for the Earth's turbulent atmosphere, will be almost as large as Keck. Astronomers hope to have it operational by 2020.

Quasar 3C 273 (the 3C stands for the Third Cambridge Catalogue) has its spectral lines redshifted in velocity by 16 percent of the speed of light. As we know, light from a source is redshifted when the object is moving away from us. In the context of Hubble's Law, such dramatic redshifts mean that quasars are receding at tremendous speeds and are very far away. The quasar 3C 273 travels at some 30,000 miles per second (48,000 km/s) and is some 2 billion light-years (640 million parsecs) distant from us.

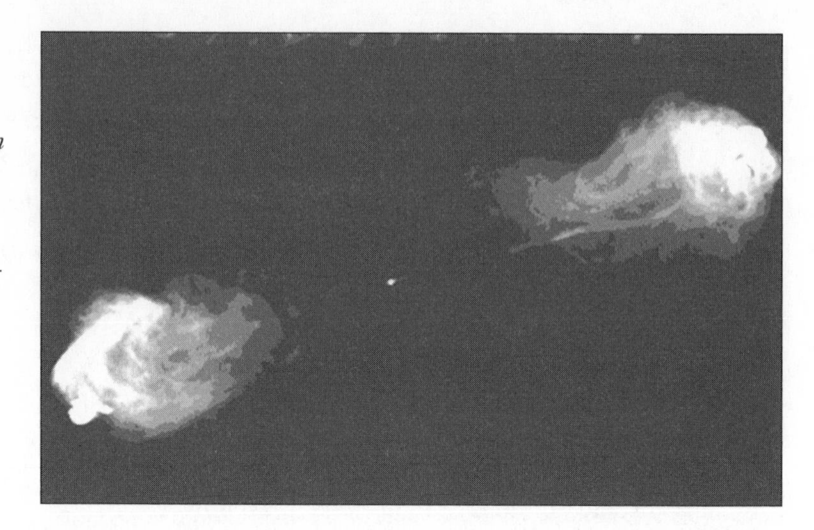

This image from the Very Large Array shows the detailed structure of Cygnus A, located in the constellation Cygnus, the Swan. One of the brightest radio sources in the sky, Cygnus A is one of a large number of radio galaxies that have been discovered with "jets" that arise from the region around a central black hole.

(Image from NRAO)

It is an awe-inspiring thing to contemplate an object so powerful and so distant. Quasars are distant and bright but also small. How do we know they are small? Many quasars flicker—rapidly varying in brightness—on scales of days. We know that light must be able to travel across the size of an object for us to see it vary in brightness. Why? Because for a region to appear to brighten, we must detect photons from the far side as well as from the near side of the object that is brightening. If an object is 1 light-year across, the "brighter" photons from the far side of the source will not reach us until a year after the photons on the near side. So this source could only flicker on scales of a year.

Astronomer's Notebook

Quasars, among the most luminous objects in the universe, have luminosities in the range of 10^{38} watts to 10^{42} watts. These numbers average out to the equivalent of 1,000 Milky Way Galaxies.

Any object that flickers on scales of mere days must be small—light-*days* across, to be exact. This characteristic flickering reveals that the source of energy is perhaps the size of our solar system—presumably a gaseous accretion disk, a collection of interstellar material, caught in the gravitational field of a supermassive black hole and spiraling toward it.

If a supermassive black hole is the source of a quasar's power, then about 10 Sunlike stars per year falling into the black hole could produce its enormous luminosity.

Quasars as Galactic Babies

Quasars might be more than strange, distant powerhouses of the early universe. In fact, they might be part of the family tree of every galaxy, including our own. Perhaps all galaxies, as they form, start out as quasars, which become less luminous, less energetic, as the early fuel supply of the galaxy is exhausted. Certainly, quasars cannot burn fuel forever at the prodigious rates that they do.

But what could fuel a quasar? A generation of stars that forms near the center of the galaxy early in its life could, through the natural mass loss that occurs as stars age, provide the needed mass. Another possibility is a galactic train wreck. If two galaxies collide, one could provide fuel for the other's dormant black hole. As the fuel arrives, the black hole once again lights up. Bingo! We have a quasar.

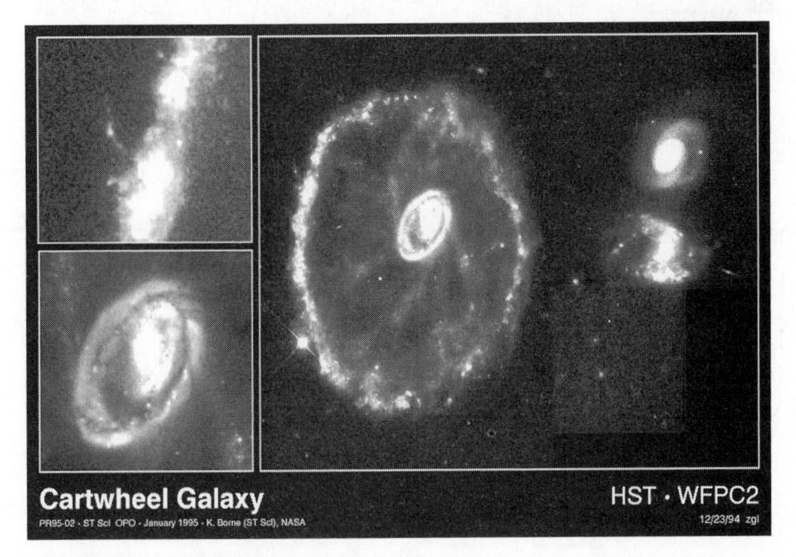

When galaxies collide, star formation is often the result, as seen in the large ring of star formation that surrounds the galaxy to the left of the two smaller galaxies. This Hubble Space Telescope image shows the famous "cartwheel" galaxy and the escaping culprit from a hit-and-run. One of the two small galaxies near the ring is the guilty party. The bright ring surrounding the cartwheel galaxy is teeming with young massive stars.

(Image from NASA)

A Piece of the Action

From what we have seen so far of galaxies, there aren't many slackers. They all seem to get a lot done in an average day. That is, they all seem quite active.

But *active galaxy* has a specific meaning to an astronomer. Indeed, what astronomers call active galaxies might well be an intermediate evolutionary stage between quasars and normal (or perhaps we should say *mature*?) galaxies.

def•i•ni•tion

Active galaxies are galaxies that have more luminous centers than normal galaxies.

Between the great distances to quasars and the more moderate distances to our local galactic neighbors are a vast number of normal galaxies and a few galaxies that are more luminous than average, particularly in their central or nuclear region. It is these latter objects that astronomers call active galaxies.

The excess luminosity tends to be concentrated in the nucleus of the galaxy, and we refer to the centers of these galaxies as active galactic nuclei. As a class, active galaxies have bright emission lines—implying that their centers are hot—that are variable on short time scales. Some of them also have jets of radio emissions emanating from their centers, stretching hundreds of thousands of light-years into intergalactic space.

The Violent Galaxies of Seyfert

In 1944, Carl K. Seyfert, an American astronomer, first described a subset of spiral galaxies characterized by a bright central region containing strong, broad emission lines. These galaxies, now called *Seyfert galaxies*, have luminosities that vary, and some show evidence of violent activity in their cores.

Seyfert galaxies have several distinguishing characteristics:

- Spectra emitted by Seyfert nuclei have broad emission lines, which indicate the presence of very hot gas or gases that are rotating at extreme velocities.
- The radiation emitted from Seyfert galaxies is most intense at infrared and radio wavelengths and must be, therefore, nonstellar in origin.
- The energy emitted by Seyfert galaxies fluctuates significantly over rather short periods of time.

Astro Byte

Only 1 percent of all spiral galaxies are classified as Seyfert galaxies. Here's another way to think about this: perhaps all spiral galaxies exhibit Seyfert properties 1 percent of the time.

At the heart of all Seyfert galaxies is something relatively small but extremely massive—massive enough to (periodically) create the tremendous activity evident at the Seyfert galaxy nucleus. One likely possibility? The core of a Seyfert galaxy, like the center of our own Galaxy, might contain a massive black hole.

Cores, Jets, and Lobes: A Radio Galaxy Anatomy Lesson

Another kind of active galaxy is called a radio galaxy. Although Seyferts are an active subclass of spiral galaxies, radio galaxies are an active subclass of elliptical galaxies.

Radio galaxies come in many types, often classified by their shapes. Some radio sources have emissions only in their nucleus. In others, two narrow streams, or jets, of oppositely directed radio emissions emerge from the galaxy nucleus. The jets in these so-called double radio sources often end in wispy, complex puffs of radio emissions much larger than the central elliptical galaxy, which are called *radio lobes.* In some radio galaxies lobe emissions dominate, and in others jet emissions dominate.

One remarkable aspect of the jets of radio emissions is that they are observed over such a huge range of scales. The jets in some galaxies are linear for hundreds of thousands of light-years, and yet are observable down to the smallest scales that we can see—a few light-years. Astronomers think radio jets are beams of ionized material ejected from near the galactic center. The jets eventually become unstable and disperse into lobes.

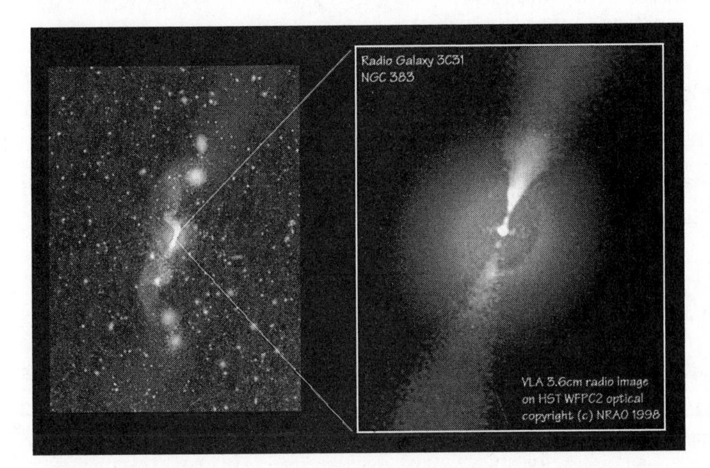

The radio galaxy 3C31 is shown superimposed on an optical image of the same region. In the inset to the right, the Very Large Array image (jets) is superimposed on a Hubble Space Telescope image of the region. When world-class telescopes like the Very Large Array and the Hubble Space Telescope improve their resolutions in coordination, useful overlays like these are possible.

(Image from NRAO/NASA)

Material in radio jets is being accelerated to enormous velocity. In some radio jets, bright blobs of radio emissions appear to be moving faster than light. But hold your horses. That's not allowed! What's going on?

The apparent *superluminal motion* (speed faster than light) results from jets that are moving toward us, giving the illusion of impossibly high velocity. They do not defy any laws of physics.

Close Encounter

Many differences in astronomy come down to a question of perspective. Astronomers are not entirely certain that core-halo radio galaxies (with a bright center and a diffuse envelope of emissions) and lobe radio galaxies (with a distinct bright center and diffuse lobes) are unique objects. That is, a core-halo galaxy might be nothing more than a foreshortened view of a lobe radio galaxy. If the galaxy happens to be oriented so that we view it through the end of one of its lobes, it will look like a core-halo galaxy. If we happen to see the galaxy from its side, we will detect two widely spaced radio lobes on either side of a central core.

What is this mysterious *radio emission* we have been discussing? Well, radio emissions come in two basic flavors. One is rather bland and the other a bit more spicy. The bland radio emission is *thermal emission*, which arises most commonly in regions of hot, ionized gas like the HII ("H two") regions around young, massive stars. This type of radio emission comes from free electrons zipping around in the hot gas.

The spicy variety is *synchrotron emission*, sometimes called *nonthermal emission.* This type of radiation arises from charged particles being accelerated by strong magnetic fields. (The nineteenth-century British physicist James Clerk Maxwell first discussed the effects of magnetic fields on charged particles.) The intensity of synchrotron radiation is not tied to the temperature of the source but to the strength of the magnetic fields that are there. Radio jets, filled with charged particles and laced with strong magnetic fields, are intense sources of *synchrotron radiation.*

Where It All Starts

Quasars and active galaxies are sources of tremendous energy. We know that stars cannot account for their energy output and that most of their energy arises from a small region at the center of the galaxy. So how can we account for the known properties of active galaxies and quasars, including their staggering luminosities?

We also know the rotation curves of many galaxies show evidence of large mass accumulations in the central regions. In fact, the masses contained are so large and in such a small volume that astronomers have in many cases concluded that a black hole must be present at the galaxy's center.

If a black hole were present in the center of active galaxies, along with the quasars that they might have evolved from, many of the observed properties of these strange galaxy types could be explained. To begin with, the tremendous nonstellar energy output originating in a compact area points to the gravitational field and accretion disk of a black hole.

But not just any black hole.

These black holes dwarf stellar-mass black holes that arise from the death of a massive star. *These* black holes are *supermassive* black holes.

The center of our own Galaxy, the Milky Way, appears to contain a black hole with more than 1 million solar masses. To account for the much greater luminosity of an active galaxy, the mass must be much higher—perhaps 1 billion solar masses.

Astronomer's Notebook

Remember the fluctuations in brightness seen in Seyfert galaxies and quasars? These fluctuations might be explained by brightness fluctuations in the accretion disk—the swirling disk of gas spiraling toward the black hole.

If black holes are active galaxy and quasar engines, we can explain the luminosity of these objects as the result of gas that spirals toward the black hole at great velocity, becoming heated in the process and producing energy in the form of electromagnetic radiation. X-ray photons are often produced by this process, and these are in fact visible with the new Chandra X-ray Observatory.

In the black hole model, the radio jet arises when the hot gas streams away from the accretion disk in the direction of least resistance—perpendicular to the accretion disk. These jets, then, can stream away from the disk in two directions, giving rise to the oppositely directed jets.

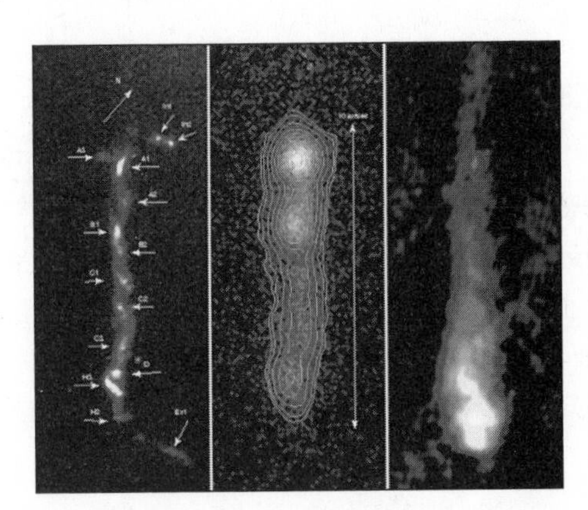

This image shows the jet associated with the quasar 3C 273. On the left is the optical image (from the Hubble Space Telescope). In the middle is the Chandra X-ray Observatory image, and on the right is the radio image of the jet as seen by the Multi-Element Radio Linked Interferometer Network (MERLIN). Notice that the different wavelengths emphasize different parts of the jet.

(Image from NASA, Chandra, and Merlin)

If black holes exist at the centers of all or most galaxies, then the differences between quasars, radio galaxies, Seyfert galaxies, and normal galaxies is not the engine but the fuel. When fuel is plentiful (as it was early in the universe), the cores of all galaxies might burn bright as a quasar. Later, they would move through a quieter, yet still active phase as Seyfert or radio galaxies. Later still, at our epoch in the universe, with fuel less plentiful, most black holes lie dormant. Models of galaxy evolution continue to be hotly debated (at least by astronomers), but the existence of central black holes might answer many questions.

Here's a closing thought, then. Let's say you take the trouble to find the quasar 3C 273 for yourself. You see a quasar—but the light that you see left the galaxy some 2 billion years ago. The quasar you see might now (now as it is experienced at the distant location of the 3C 273) be a more mature galaxy, not unlike the Milky Way. Perhaps on some planet orbiting an average star somewhere in the distant universe, an amateur astronomer is pointing a telescope at our Milky Way, which appears to him or her or it as a faint, bluish blob of light: a quasar.

The Least You Need to Know

- Quasars were first discovered at radio wavelengths and are some of the most distant astronomical objects visible.
- Quasars might be the ancestors of all galaxies, the violent beginnings of us all.
- Active galaxies are any galaxies that are more luminous than what we call normal galaxies, with bright starlike cores and broad, strong emission lines.
- Seyfert galaxies are the active subset of spiral galaxies, and radio galaxies are the active subset of elliptical galaxies.
- Radio jets originate at the cores of active elliptical galaxies and terminate in wispy patches of emissions called lobes.
- Supermassive black holes are the most likely source of energy for quasars and active galaxies.

Chapter 16

Cosmology and Cosmologies

In This Chapter

- The job description of a licensed cosmologist (hint: it doesn't involve makeup)
- Understanding the cosmological principle
- All about the Big Bang
- Critical density and the expanding universe
- Matter from energy and energy from matter

Cosmology. The word has an archaic ring to it. And that's no wonder because cosmology has been part of the English language since at least the 1600s and describes the study of some of the oldest and most profound issues humankind has ever addressed: the nature, structure, origin, and fate of the universe. Those are some big questions!

At its root is the word the Greek philosopher Pythagoras used to describe the universe, *kosmos*, and for centuries, cosmology was the province of philosophers and priests. In this chapter, we look at what happened when astronomers tackled the subject in the twentieth and twenty-first centuries.

The Work of the Cosmologist

Many of us grew up thinking that the universe was both infinite in size and eternal in time. The planets might move in the sky, and, farther away, stars might orbit the centers of galaxies, but on the largest scales, certainly the universe must be changeless. It has always been and it will always be. How, after all, could it be otherwise?

This was just one possibility to consider as modern cosmologists began to study the universe. Perhaps it never had a beginning, but of course, virtually every culture, mythology, and religion has thought otherwise. Civilization abounds with creation stories—mythological and religious narratives that explained the origin of the universe. In the Judeo-Christian tradition, for example, God said, "Let there be light." And there it was. Muslims believe that, in the time before time, there was only God, who, when He wanted to create something, needed only pronounce the verb *be*. So it was that the world and the heavens came into being. In Hinduism there are diverse beliefs about the origin of the universe, but many believe the universe cyclically recreates itself once all *karma* is extinguished. Some Buddhists believe the world recreates itself every tiny fraction of a second.

Astronomer's Notebook

For decades, astronomers had to struggle with cosmological questions based on a bare minimum of observational evidence. The past two decades have provided observers and theorists with a wealth of new information about what the universe looked like in its infancy. These observational clues have triggered an explosion in our understanding of the origin of the universe.

The human mind naturally looks for origins, beginnings, and grand openings, and in the twentieth century, scientific cosmologists stumbled across two important bits of evidence of the biggest grand opening ever, evidence that there had, in fact, been a beginning.

Modern cosmologists have put together a creation story of their own. It is the result of over a century of observation, modeling, and testing. The model is so well established that it is generally referred to as the *standard model*. As Steven Weinberg points out in his account of the start of it all, *The First Three Minutes* (Basic Books, 1993), the study of the origins of the universe was not always a scientifically respectable pursuit. Only in relatively recent scientific history have we had observational tools rigorous enough to test theories of how the universe began.

Two New Clues

The twentieth century threw some major curve balls at the human psyche, from the carnage of two world wars to the invention of the hydrogen bomb, capable of releasing the fury of a stellar interior on the surface of our planet. And although the seventeenth century saw Earth pushed from the center of the known universe, it was early twentieth-century astronomers who told us that we were located in the suburbs of an average galaxy, hurtling away from all the other galaxies that we can see.

Why are we all flying away from one another? Because the universe exploded into being about 14 billion years ago!

Redshifting Away

From our study of Hubble's Law and Hubble's constant, we know by observing the redshift in the spectral lines of galaxies that all the galaxies in the universe are moving away from us and those farthest away are moving the fastest. We also know we are not in a special location and any observer in any galaxy would see exactly what we do.

Armed with Hubble's Law and a value for the Hubble constant, we can easily calculate how long it has taken any given galaxy to reach its present distance from us. In a universe that is expanding at a constant rate, we simply divide the distance by the velocity. The answer for all the galaxies we see is about 14 billion years.

So if we were able to rewind the expanding universe to see how long it would take for all the galaxies that are flying apart to come together, the answer would be close to 14 billion years.

Pigeon Droppings and the Big Bang

The Russian-born American nuclear physicist and cosmologist George Gamow (1904–1968) was the first to propose, in the 1940s, that the recession of galaxies, which we've known about since the 1920s, implied that the universe began in a spectacular explosion. This was not an explosion *in* space but an explosion *of all* space. The explosion that started the universe filled the entire universe; that is, at the instant of creation, the universe *was* an explosion.

Later, British mathematician and astronomer Sir Fred Hoyle (1915–2001) coined the term *Big Bang* for this event. And he did not intend it as a compliment. If the phrase sounds silly and trivial, it's because Hoyle thought the idea of a primordial explosion was something like that, silly and trivial. Therefore, with astronomers Herman Bondi and Thomas Gold, Hoyle proposed an alternative, what he called the *steady state theory* of the universe, which holds that the universe is eternal, without beginning and without end.

Despite Hoyle's derisive intentions, the name "Big Bang" stuck, and quite soon after Gamow put forth the theory, other theorists realized that if there were an origin as proposed, the early universe must have been incredibly hot and the electromagnetic echoes of this colossal explosion should therefore be detectable as long-wavelength radiation that would fill all space. This radiation was called the *cosmic microwave background.* In short, this was one creation myth that they could test.

The science of radio astronomy was the accidental by-product of a telephone engineer's search for sources of annoying radio interference. Another radio-frequency accident resulted in an even more profound discovery. From 1964 to 1965, two Bell Telephone Laboratories scientists, Arno Penzias and Robert Wilson, were working on a pioneering project to relay telephone calls via satellites. Wherever they directed their antenna on the sky, however, they detected the same faint background noise. The noise was not associated with any particular point on the sky but was the same everywhere. In an effort to remove all possible sources of interference, they went so far as to scrape off the pigeon droppings they found on the antenna because these were a potential source of heat—energy, which could be a source of radio interference. But even when their antenna was clean as a whistle, the noise persisted.

The "noise" they had stumbled across was the very cosmic microwave background that had been predicted, the highly redshifted photons left behind by the Big Bang. The theory goes like this: immediately after the Big Bang, the energy liberated was most intense in the shortest, highest-energy wavelengths—gamma rays. As the universe expanded and cooled over millions, then billions of years, the waves expanded as well, reaching us today as long-wavelength photons in the microwave part of the electromagnetic spectrum.

Astro Byte

Astronomers John C. Mather and George Smoot won the Nobel Prize in Physics in 2006 for their work on the cosmic background radiation. For the first time, the data mapped tiny ripples in the background radiation that existed only a few hundred thousand years after the Big Bang.

This microwave radiation carries information from the early moments of our universe. It is the remnant echo of the Big Bang permeating the universe, evidence that in past times the universe was a much hotter place than it is now. To study this echo in detail, scientists launched the Cosmic Microwave

Background Explorer (*COBE*) satellite in November 1989. In the first few minutes of operation, it measured the temperature of the background radiation to be 2.735 K.

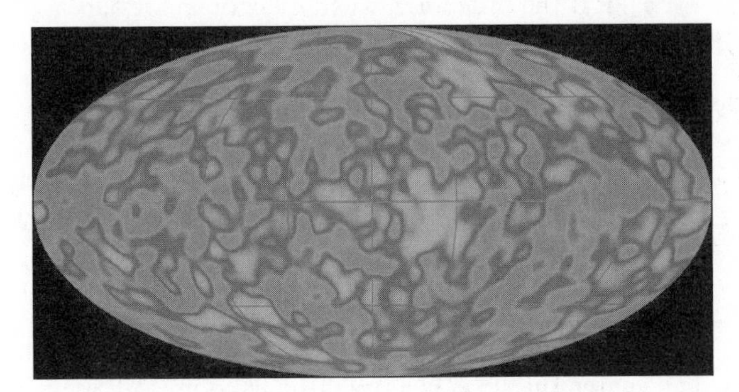

The COBE *Differential Microwave Radiometer (DMR) found variations in the cosmic microwave background (shown here as brightness variations) above and below a temperature of 2.73 K. The variations are tiny—at the level of a few parts in a million.*

*(Image from NASA/*COBE*)*

The data from this satellite enabled scientists to make great improvements on the early, crude mappings of the cosmic microwave background. In addition, we have also learned from *COBE* that the solar system has a small motion relative to the cosmic microwave background. Finally, in 1992, *COBE* data revealed yet another exciting fact: there are tiny fluctuations on the sky in this background radiation; it is not exactly homogeneous or smooth. These temperature differences are the result of density fluctuations in the early universe, which appear to have seeded the universe for the eventual formation of galaxies.

Same Old Same Old

Edwin Hubble also made two very important discoveries in his studies of galaxy types and distributions. He found that the universe appeared to be both isotropic (the same in all directions) and homogeneous (one volume of space is much like any other volume of space).

His conclusion was the result of painstaking study of the distribution and type of a large number of galaxies at various redshifts or distances from Earth.

Together, the homogeneity and isotropy of the universe make up the *cosmological principle*, one of the cornerstone assumptions of modern cosmology. If we could not make this assumption—which is based on observation—then our cosmology might only apply to a very local part of the universe, but the cosmological principle enables us to extrapolate conclusions drawn from our local viewpoint to the whole universe.

def•i•ni•tion

The **cosmological principle** is a cornerstone assumption (based on observation) about the nature of the universe. It holds that the universe exhibits two key properties: homogeneity (sameness of structure on the largest scales) and isotropy (it looks the same in all directions).

Consider these important implications: a homogeneous universe can have no border or edge because it is the same in any volume, nor can it have a center because it should look the same in all directions from any viewpoint.

Earlier, we introduced the most familiar model of the expanding universe: a toy balloon, on which we randomly drew some dots with a magic marker to represent galaxies.

If you, as a two-dimensional being, look out into your two-dimensional universe—that is, out into the balloon's surface—you see no center and no edge, and if the balloon is inflated, the dots move away from one another. Regardless of their point of view, all the dots "see" all the others as moving away.

Of course, the balloon model has its limitations. If the universe is represented by the surface of a balloon, then we are talking strictly about a two-dimensional universe. If you can draw only on the surface, there is effectively no inside of the balloon. Nevertheless, a two-dimensional model of our three-dimensional universe helps us picture the expansion. Another point to remember is that the balloon model represents a "closed" universe, which we discuss in the next chapter. In an infinite and "flat" universe, we represented the universe with an infinitely large rubber sheet, which we stretched in all directions.

Big Bang in a Nutshell

You might be sitting there thinking, "But what caused the Big Bang?"

That is a question mostly outside the bounds of astronomy and physics. Scientists can only hope to understand things of this universe, and whatever caused the Big Bang is inherently unobservable.

Universal recession in time means the universe is expanding, and expansion, in turn, implies an origin time. Astronomers believe that at its origin, the universe was unimaginably dense. The entire universe was a point, with nothing outside the point. About 14 billion years ago, the entire universe exploded in the Big Bang, and the universe has been expanding ever since, cooling and coalescing into ever more organized states of matter. One very highly organized state of matter (you) is reading this book. Hubble's Law describes the rate of the observed expansion.

Big Bang Chronology

The prehistory of the Big Bang is inherently unknowable, but we are able to discuss what happened very soon *after* the Big Bang. And we do mean *very* soon. From about $1/100$ of a second after the Big Bang onward, we can outline the major steps in the evolution of the universe, which is basically a story of cooling and expanding. For a thorough review of modern cosmology, see Timothy Ferris's excellent and humorous *The Whole Shebang* (Touchstone, 1998). And you can find an expanded and more technical take on the early moments of the universe in Steven Weinberg's *The First Three Minutes* (Basic Books, 1993).

At the earliest times we can track, the universe was incredibly hot (10^{11} K) and filled with elementary particles, the building blocks of atoms: electrons, positrons, neutrinos, and photons of light. And for every billion or so electrons present at this time, there was one heavy particle (a neutron or proton).

In this roiling soup—an entire universe hotter than the core of a star—energy and matter were going back and forth as electrons and positrons *annihilated* to produce energy, and more were born from the energetic photons that filled the universe. At about 1 second into the life of the universe, the temperature everywhere was 10^{10} K, still too hot for neutrons and protons to be bound into nuclei.

def•i•ni•tion

Particles and antiparticles **annihilate** when they meet, converting their mass into energy. Electrons and anti-electrons (positrons) were continually created and annihilated in the early universe.

As the universe rapidly cooled, after a mere 10 to 15 seconds, there was insufficient energy to create new electrons and positrons, so most of them annihilated, without new ones taking their place. It was cool enough at this point for helium nuclei (two protons and two neutrons) to form, but they could only form after deuterium (an isotope of hydrogen consisting of one proton and one neutron) formed. The instability of deuterium at this temperature means that the universe still couldn't form much helium.

After about the first three minutes of its existence, the universe had cooled to some 1 billion degrees. Most of the electrons and positrons had annihilated, and the universe then consisted of photons of light, neutrinos and antineutrinos, and a relatively small amount of nuclear material. At about this time, deuterium nuclei were able to hold together, and the entire universe (for a moment) acted like the core of a star, fusing helium and small amounts of lithium in a phase called *primordial nucleosynthesis.*

The universe was still far too hot for electrons to come together with nuclei to form stable atoms. Not until about 300,000 years later (when the entire universe was at the temperature of a stellar photosphere) would nuclei be able to hold on to electrons. It wasn't until after this first 300,000 years that the universe became transparent to its own radiation.

What does that mean?

Before the electrons settled out into atoms, they got in the way of all the photons in the universe and kept colliding with them. With the electrons cleared out of the picture—grabbed by nuclei—the universe shifted from a *radiation-dominated* state to a *matter-dominated* state.

Here's another way to think about this idea: for the first 300,000 years after the Big Bang, the universe was "foggy" with electrons. This electron "fog" made the universe opaque to its own photons. When the universe was cool enough for electrons and nuclei to combine, it suddenly became transparent. The cosmic microwave background comes from those first freed photons.

From this point on, the universe continued to cool and coalesce, eventually forming the stars and galaxies of the observable universe.

A Long Way from Nowhere

Now just *where* did this Big Bang occur, you might be wondering?

If the Big Bang took place *somewhere*, that place is, by definition, the center of the universe. Yet the cosmological principle forbids any such center. The Big Bang might resolve some open questions, but it also torpedoes the cosmological principle.

Astro Byte

At the moment of the Big Bang, the entire universe came into being. As the universe cooled, the elements hydrogen (73 percent) and helium (27 percent) were created in the primordial fireball, along with small amounts of lithium.

Well, actually, it doesn't—as long as we conclude that the Big Bang took place *everywhere*. And the only way such a conclusion could be true is if the Big Bang was not an explosion within an empty universe—like a movie explosion, say—but was an explosion of the universe itself. At early times in the universe, its entire volume was hotter than a stellar interior. The universe was and is all that exists and has ever existed.

Looked at this way, it is a symptom of geocentric bias to say that the galaxies are receding from us. Rather, the universe itself is expanding, and the galaxies, ourselves included, are moving along with that expansion.

How Was the Universe Made?

Radiation dominated over matter in the early universe. Although photons (radiation) still outnumber atoms (matter) by about a billion to one in the universe, matter now contains far more energy than radiation. A simple calculation, using Einstein's celebrated equation showing the equivalence of energy and mass, $E = mc^2$, demonstrates that matter, not radiation, is dominant in the current universe. The energy contained in all the mass in the universe is greater than the energy contained in all the radiation.

There is a twist here, however, which we explore in the next chapter. The entire universe might be dominated by "dark energy" driving its continued expansion.

As the universe expands, both radiation and matter become less concentrated; they travel with the expansion. However, the energy of the radiation is diminished more rapidly than the density of matter because the photons are redshifted, becoming less energetic as their wavelength lengthens. Thus, over time, matter has come to dominate the universe.

Mommy, Where Do Atoms Come From?

When we looked at fusion in the cores of stars, we noted that hydrogen served as the nuclear fuel in a process that produced helium. Yet so much helium is in the universe—it accounts for more than 23 percent of the mass of the universe—that stellar fusion cannot have produced it all. It turns out that most of the helium in the universe was created in the moments following the Big Bang, when the universe had cooled sufficiently for nuclei to hold together, after about the first three minutes.

Hydrogen fusion in the early universe proceeded much as it does in the cores of stars. A proton and a neutron come together to form deuterium (an isotope of hydrogen), and two deuterium atoms combine to form a helium nucleus.

Astro Byte

When the entire universe was the temperature of a cool stellar photosphere (about 5,000 K), electrons were able to join together with nuclei, forming hydrogen and helium atoms and causing the universe to become transparent to its own radiation.

Careful calculations show that between 100 and 120 seconds after the Big Bang, deuterium would be torn apart by gamma rays as soon as it was formed. However, after about two or three minutes, the universe had sufficiently cooled to allow the deuterium to remain intact long enough to be converted into helium. As the universe continued to cool, temperatures fell below the critical temperature required for fusion, and primordial nucleosynthesis ended.

Only after about 300,000 years had passed would the universe expand and cool sufficiently to allow electrons and nuclei to combine into atoms of hydrogen and helium. At this point, with the temperature of the entire universe at some 5,000 K, photons could first move freely through the universe.

The cosmic microwave background we can now measure consists of photons that became free to move into the universe at this early time.

The rest of the periodic table—the other elements of the universe—beyond lithium would be filled out by fusion reactions in the cores of stars and supernova explosions of massive stars.

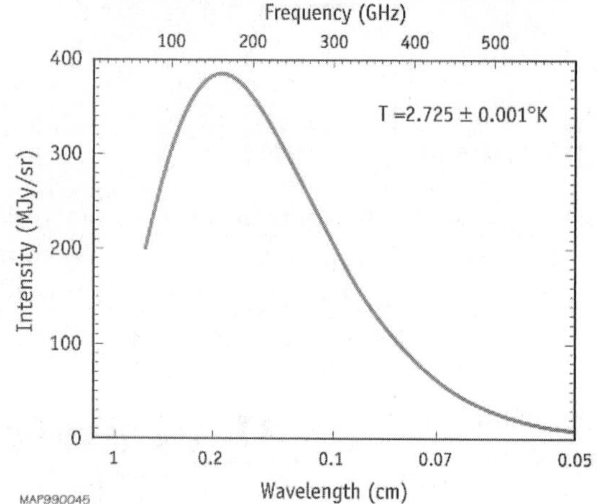

The cosmic microwave background (CMB) can be very well modeled with a black body that has a temperature of 2.725 K (curve shown here).

(Image from NASA)

Stretching the Waves

As mentioned earlier, there is another way to think about the observed redshift in galaxy spectra. The photons emitted by a receding galaxy are like elastic bands in the fabric of the universe: they expand with it. In this sense, the redshift we measure from distant galaxies is nothing less than a *direct* measurement of the expansion of the universe.

We have seen where the universe and all that is within it came from. Now we turn finally to where it is all going and where it will end. Will the universe expand forever? Or will it turn back in on itself and end in a final collapse? It all depends on gravity.

Depending on how much mass there is, the universe will either expand forever or collapse back in on itself in the opposite of a Big Bang: a Big Crunch. Recent observations of high-redshift supernovae and the cosmic microwave background suggest this Big Crunch will never happen. The universe appears destined—or doomed—to expand forever.

The Least You Need to Know

- The redshift of galaxies and the presence of a cosmic microwave background are two clues that the universe had a beginning in time.
- The cosmological principle, which allows us to generalize from local observations, holds that the universe is homogeneous (uniform in structure on large scales) and isotropic (it looks the same in all directions).
- The current theory of the origin of the universe is the Big Bang, which holds that the universe began as an explosion that filled all space.
- For the first 300,000 years after the Big Bang, the universe was "foggy" with electrons and opaque to its own photons, but when the universe was cool enough for electrons and nuclei to combine, it became transparent, and the cosmic microwave background comes from these first freed photons.
- All matter in the universe was created at the moment of the Big Bang, and since that time, stellar fusion and supernovae have converted the dominant hydrogen and helium into all of the elements of the periodic table.

Chapter 17

The Beginning and the End of the Universe

In This Chapter

- Supernova light curves as standard candles
- The accelerating expansion of the universe
- The scale of fluctuations in the cosmic microwave background (CMB)
- The future of the universe: expansion or contraction?
- The density of the universe may determine its fate

In Woody Allen's 1977 movie *Annie Hall*, a flashback takes us back to the Brooklyn childhood of the main character, a stand-up comic named Alvy Singer. The family doctor has been summoned because young Alvy, perhaps 10 years old, is depressed. Having just learned that the universe is expanding, he believes it will eventually fly apart. The physician, puffing on a cigarette, offers the comforting thought that universal calamity is many billions of years in the future. Alvy's mother is more strident: "What is it your business? We live in *Brooklyn*. Brooklyn is *not* expanding!"

What Redshift Means

Alvy's mother was right. The truth is that Brooklyn itself *isn't* expanding. But the universe that revolves around it (we're taking a Brooklynite's perspective here) is expanding. Much to the chagrin of New Yorkers, Brooklyn is not somehow exempt from the laws of the universe. This borough—like Earth, like a star, like a galaxy, or like a human being—is held together by its own internal forces. The expansion of the universe that is the echo of the Big Bang is only visible on the largest scale, the distances between clusters and superclusters of galaxies.

Let's return to our description of the universe as the surface of a rubber balloon. As the balloon is inflated, the distance between any two dots drawn on its surface increases. The resulting observation is that these receding galaxies—they are receding with respect to one another—emit photons that are redshifted. Another way to think of it is that the photons that leave another galaxy are part of the fabric of the universe, and as this fabric is stretched, each photon gets stretched along with it—to a longer, and therefore redder, wavelength. This type of wavelength increase is called a *cosmological redshift* and is a direct measure of universal expansion.

def•i•ni•tion

Cosmological redshift is the lengthening of the wavelengths of electromagnetic radiation caused by the expansion of the universe.

Here Are Your Choices

Please keep this in mind: the values we measure locally, related to the expansion of the universe, are not necessarily "universal" in time. Just because the universe is expanding at a certain rate now in our particular neighborhood doesn't mean that it has always expanded at that rate. It might have been expanding more slowly or more quickly in the distant past. The only way to find out is to measure the redshift of very distant objects.

And just because the universe is expanding now doesn't necessarily mean that it will go on expanding forever. The expansion that presumably started with the Big Bang is opposed by a relentless force you have met many times before in this book: gravity. The eventual fate of the universe hangs on how much stuff (mass and energy) there is in the universe. The density of matter and energy in the universe will determine how it all ends.

We have just three choices: the universe will either expand forever (be "unbound"); it will continually slow in its expansion but never stop (be "marginally bound"); or it will reach a certain size and then begin to contract (be "bound").

A Matter of Density

The origin of these limited choices is the density of the universe. If the universe contains a sufficient amount of matter and energy—if it is sufficiently dense—then the force of gravity will be such that it will eventually succeed in halting expansion. The universe will gradually expand at a slower and slower rate, and then reverse itself and begin to collapse. If, however, the universe (even counting all of the material that we can't see—the "dark matter") does not contain sufficient mass—is too thin—then the expansion of the universe will continue forever.

Remember that in the Big Bang theory, there is no such thing as new or old matter. All matter was created as a result of the Big Bang; therefore, the density of the universe—and its eventual fate—were determined at the instant the Big Bang occurred.

A certain density of matter and energy in the universe is the sword edge between ultimate collapse and eternal expansion. And astronomers have calculated this critical density. If the universe is more dense than this, it will eventually collapse; if it is less dense than this, it will expand forever; and if it is exactly this dense, it will still keep expanding, but more and more slowly, forever.

Stalking the Wily Neutrino

We've seen neutrinos before, streaming from the Sun. These are subatomic particles without electric charge and with virtually no mass. As a result, they are sort of stealth particles: small, neutral particles that are hard to detect. On June 5, 1998, U.S. and Japanese researchers in Takayama, Japan, announced that, using an enormous detection device, they found evidence of mass in the neutrino. The device, called the Super-Kamiokande, is a tank filled with 12.5 million gallons of pure water and buried deep in an old zinc mine.

Because neutrinos lack an electric charge and rarely interact with other particles, passing unnoticed through virtually any kind or thickness of matter, they have proven especially elusive. They are the greased pigs of the subatomic world, so plentiful that 100 billion of them (that's as many stars as there are in our Galaxy) pass through your body every second. The Super-Kamiokande contains so much water, however, that occasionally a neutrino does collide with another particle and produces an instantaneous flash of light, which is recorded by a vast array of light-amplifying detectors.

So if the researchers are right and neutrinos do have mass, just how much mass do they have? Well, it doesn't appear to be enough to affect critical density.

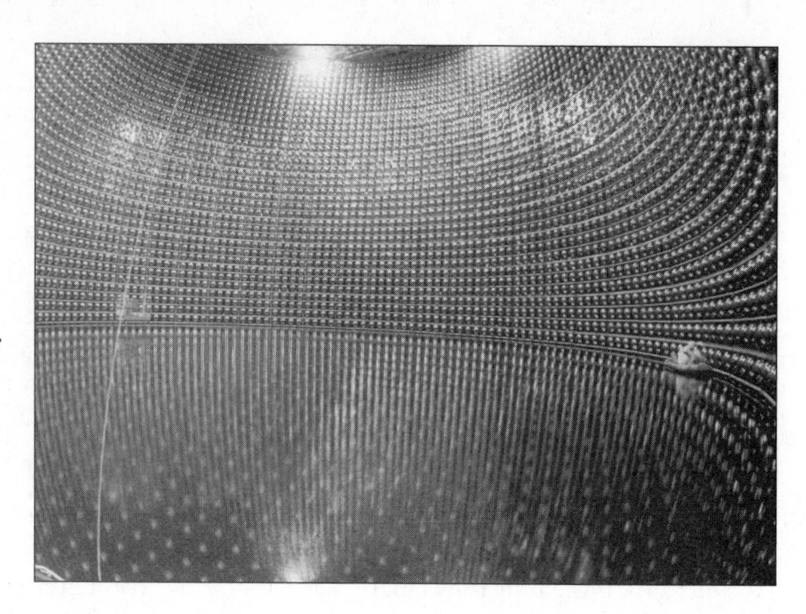

The Super-Kamiokande detector, which began observations in 1996, consists of a total of 50,000 tons of water. There are outer and inner volumes, the inner volume ringed with 11,200 photomultiplier tubes (PMTs), seen here during maintenance. The tubes are sensitive to light called "Cerenkov light," which is emitted by high velocity particles traveling in the water.

(Image from U. Tokyo/UC Irvine/BU)

Run Away! Run Away!

To predict whether the universe will expand infinitely, it is necessary to calculate critical density and to see whether the density of our universe is above or below this critical figure. By measuring the average mass of galaxies within a known volume of space, astronomers derive a density of luminous matter well short of critical density, approximately 10^{-28} kg/m^3. That figure is roughly 1 percent of what is required to "close" the universe. After factoring in the added contribution of the dark matter that has an apparent gravitational effect in galaxy clusters, we can account for about 30 percent of the critical density. Is the universe, then, far from critical density?

On the contrary, recent experiments strongly suggest that the universe is very close to critical density and that the "missing" 70 percent is not made up of mass but of energy. If the expansion of the universe is accelerating and the universe has "critical density," these might both be explained by energy contained in the vacuum of space.

What Does It All Mean?

You should now start to see the outlines of the different possible fates for the universe. Let's consider what some of the various ends of the universe might mean for us.

The Universe: Closed, Open, or Flat?

So then the universe expanded from this point in the Big Bang. What was the eventual result? What is the geometry of the universe?

To begin to visualize the shape of the universe, we have to stop thinking like Newton and start thinking like Einstein. Remember, Einstein thought of gravitation not as a force that objects exert upon one another but as the result of the distortion in space that mass causes. Einstein explained that the presence of mass warps the space in its vicinity. The more mass—the greater the density of matter—the greater the warp (the more space curves).

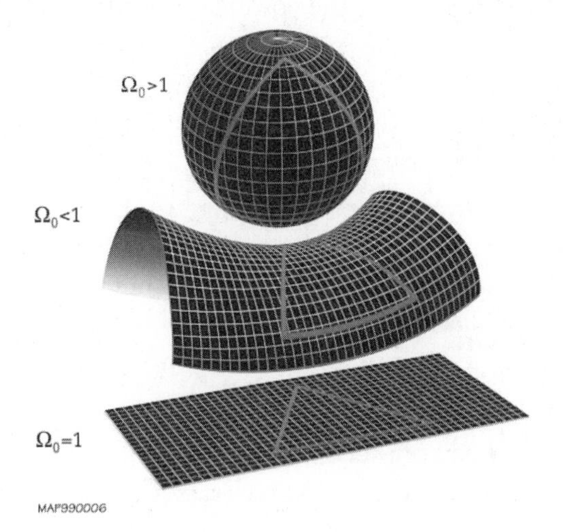

The universe might be "closed," "open," or "flat," depending on whether the universe is at critical density or not. This image shows two-dimensional models that we can use to understand these three basic geometries. A closed universe (top figure) can be thought of as curved like the surface of a sphere, and finite in extent. An "open" universe (middle figure) can be thought of as saddle-shaped and infinite in extent. A "flat" universe (bottom figure) can be thought of as a plane, also infinite in extent.

(Image from NASA)

If the density of the universe is greater than the level of critical density ($\Omega_0>1$), then the universe will warp (or curve) back on itself, closed and finite. If the universe is closed, then our balloon analogy has been particularly accurate. In a closed universe, we think of the entire universe as represented by the surface of a sphere, finite in extent, but with no boundary or edge. If you shoot out parallel beams of light in a closed universe, they would eventually cross paths, like lines of longitude on a globe.

If the density of the universe is below the critical level ($\Omega_0<1$), an open universe will result. Whereas the spherical or closed universe is said to be positively curved, the open universe is negatively curved.

Saddle Up the Horses: Into the Wide-Open Universe

The closest two-dimensional approximation of the open universe is the surface of a saddle (or a Pringles potato chip), curved up in one direction and down in the other—and extending out to infinity. In an open universe, beams of light initially parallel would diverge. Until recently, this appeared to be the type of universe we inhabited.

With precisely critical density, by the way, the universe becomes flat, albeit still infinite in extent. In what we call flat space, parallel beams of light would remain parallel forever. For this reason, flat space is sometimes called Euclidean space. Most observational results from actual measurements of the cosmic microwave background suggest that the universe might well be flat.

We Have Some Problems

Scientists are to their theories as overbearing parents are to their children. They have only the very highest expectations for them. Contradictions are not allowed and must be resolved. Even something considerably short of an outright contradiction is intolerable. A good theory should account for all observations and make testable predictions. Any observations in contradiction to the theory are taken seriously and the theory amended, if necessary.

In this regard, although it has been very successful, the Big Bang theory seems to fall short in two areas. It does not explain the incredible "sameness" of the universe on its largest scales, nor does it account for why the universe has a density, roughly 30 percent mass and 70 percent energy, that is apparently so close to critical.

With regard to the sameness issue, there is the so-called horizon problem, which deals with the incredible uniformity of the cosmic microwave background. No matter where we look, its intensity is the same, even in parts of the universe that are far too distant to be in contact with one another. Regions in the universe that never could have exchanged information—because of the limiting speed of light—seem to know about each other.

This uniformity does not contradict anything in the Big Bang theory, it's just that nothing in the theory accounts for it. The theory provides no particular reason why two widely separated regions should be the same—especially because regions very distant from one another could never have interacted. (That is, given the age of the universe, the speed of light isn't fast enough for information to travel between the most distant regions.)

With regard to the second issue, the approximation of critical density, there is the so-called flatness problem. Its name comes from the fact that the critical density of the universe, once the mass of dark matter and the role of energy are figured into things, approaches 1, which means that the universe is almost flat. Why is this a "problem"?

Again, as with the horizon problem, the difficulty is not that flatness contradicts the Big Bang theory—it does no such thing—but the theory doesn't explain *why* the universe should have formed so close to critical density. As far as the theory goes, the universe might have been significantly more dense than 1 or significantly less. Scientists won't accept the luck of the draw. They want an explanation.

Close Encounter

Four known forces govern the universe: gravity, electromagnetism, the weak nuclear force, and the strong nuclear force. Gravity and the electromagnetic force act over large distances—the size of the universe—and govern the motions of planets and galaxies, molecules and atoms. The nuclear forces act over only small distances (within an atomic nucleus) and hold nuclei together. The nuclear forces are the strongest, but they act over the shortest distances. Gravity is by far the weakest, but its long reach means that its pull will govern the fate of the universe.

In the early universe, scientists believe these forces started as a single force. As the universe expanded, each force established its own unique identity.

Down to Earth

Successful people in all fields—from astronomy to retail sales—learn to see problems not as obstacles, but as opportunities. The biggest, most basic issues astronomy deals with seem the furthest removed from our everyday experience; but the possible resolution of the horizon and flatness problems might help us to connect more intuitively with this strange thing called the universe.

Blow It Up

Within the first infinitesimal fractions of the first second following the Big Bang, astronomers have theorized, the three forces other than gravity in the universe—electromagnetic, weak nuclear forces, and strong nuclear forces—were united as a single force.

Between 10^{-35} seconds and 10^{-32} seconds after the Big Bang, gravity had already split out as a separate force, but the other forces were still one. For an unimaginably brief instant, gravity pushed the universe apart instead of pulling it together. Theorists call this moment *inflation*, for obvious reasons. Within this *inflationary epoch*, the universe expanded 100 trillion trillion trillion trillion (10^{50}) times.

Of course, 10^{-35} to 10^{-32} seconds seems to us instantaneous; but there was time—however brief—before this, between the Big Bang and the onset of the epoch of inflation. During this period, all parts of the universe were in communication with one another. They had ample time to establish uniform physical properties *before* the epoch of inflation pushed them to opposite sides of the universe.

As the early universe continued to expand, it did so at a rate faster than its constituent regions could communicate with each other. The universe, in effect, outran information, so that the most extreme regions have been out of communication with one another since 10^{-32} seconds after the Big Bang.

Yet they share the properties they had at the very instant of creation and continue to share these properties today. Thus, with the addition of inflation, the Big Bang can account for the horizon problem, and we've tied up one loose end of the theory.

Looks Flat to Me

That leaves us with the flatness problem. Why—in terms of the Big Bang theory—should the universe be at or near critical density?

If a mass of external evidence didn't exist, it would be almost impossible to convince anyone that the Earth is round. After all, it certainly *looks* flat.

After we accept that Earth is curved, however, we understand that it looks flat because the radius of the globe is very large and, therefore, the arc of the curve it describes is extremely gradual. Take a small portion of any arc, and it will look, for all practical purposes, like a straight line—that is, flat. If Earth were very small—or we were very large or very far away—the curvature of Earth's surface would be apparent on a routine basis.

The universe might have been curved at the instant of its creation, but in expanding 10^{50} times during the epoch of inflation, it became—on the scale of the observable universe—flat.

A flat universe is consistent with a universe whose density is exactly critical. If the universe were more dense, it would be positively curved; if less dense, it would be negatively curved. If the universe is truly flat, its density must be exactly critical.

I Thought We Were Done

You know that feeling when you've gotten just about everything figured out and are ready to kick back and relax for a while—but then you notice something that suddenly makes you realize that you still have a lot of work to do?

For a few years there in the 1990s, it was starting to seem like many things about the origin of the universe, the Big Bang, and the Hubble Law were fitting together very nicely.

And then, as is often the case in astronomy, several surprising new observations were made that forced astronomers to think about the workings of the universe in a completely new way.

The Universe: Smooth or Crunchy?

One of our early assumptions about the universe is that it is homogeneous and isotropic. This means that the universe appears to be the same over large volumes (homogeneous) and looks the same in all directions (isotropic). Launched in 1989, the Cosmic Microwave Background Explorer (*COBE*) satellite made observations of the cosmic microwave background (CMB) and confirmed that the universe was indeed isotropic. To very high accuracy, after the motion of the solar system through the CMB was subtracted, the CMB was incredibly smooth.

Or was it? In 1992, after more analysis of the data, the *COBE* team announced a surprising result. At very low levels—levels of 1 part in 100,000—fluctuations *did* exist in the CMB after all. That is, the CMB appeared to have some structure. At very low levels, the universe was not isotropic!

Small Fluctuations

As the first instrument to measure the CMB from space, *COBE* certainly had some limitations, but there was no doubt that it could accurately measure the microwave temperature. Using the Differential Microwave Radiometer (DMR), the *COBE* team measured the "sky temperature" with amazing accuracy and found it to be 2.725 K (give or take 0.001 K).

What *COBE* could not do was map the temperature variations with high resolution. The *COBE* image had an effective resolution of about 10 degrees on the sky. Remember that the Moon is about a degree on the sky, so the fluctuations were blurred to the size of 20 Moons.

This image from COBE *(the Cosmic Microwave Background Explorer) shows three views of the all-sky map of the cosmic microwave background (CMB). The top image shows combined emissions from the Milky Way and our motion through the CMB. The middle image has had our motion (the dipole) removed and shows only our Galaxy's emissions. The map showing only small-scale fluctuations in the CMB is at the bottom.*

(Image from COBE/NASA)

But the fact that there were fluctuations at all was very interesting to cosmologists. It meant that at a very early time in the universe (only about 300,000 years after the Big Bang), there were density variations in the universe at the level of 1 part in 100,000. It was quite possible that these density variations were the "seeds" that would eventually grow, thanks to gravity, into galaxies, galaxy clusters, and galaxy superclusters. The astronomers responsible for this finding, John Mather and George Smoot, won the Nobel Prize in 2006.

Cosmologists had often asked how, if the universe were indeed perfectly homogeneous, could stars, galaxies, or anything else form? Now they had a partial answer. The universe was not perfectly homogeneous. Something created density variations in the early universe. The CMB betrayed the presence of a little "crunchiness" in what had been thought to be a "smooth" early universe.

Zooming In

The *BOOMERANG* (Balloon Observations Of Millimetric Extragalactic Radiation And Geophysics) mission flew for 10 days in late 1998 and early 1999. *BOOMERANG* did not try to compete with the *COBE* mission by mapping the whole sky, nor did it try to compete with NASA's then upcoming *MAP* (Microwave Anisotropy Probe) mission by mapping the whole sky at high resolution. Instead, the balloon-borne craft flew 37 kilometers above Antarctica, sampling high-resolution data at microwave frequencies, sensitive to the part of the spectrum where the universe emits photons that are a remnant echo of the Big Bang.

BOOMERANG mapped only a tiny portion of the sky, but it did so with a much higher resolution than *COBE*. It made images at four frequencies, and all four frequencies were imaged with less than a degree resolution, more than 10 times "sharper" than the *COBE* images.

Time for Some Geometry

The results were nothing short of breathtaking. *BOOMERANG* was able to detect the true-scale size of the fluctuations in the cosmic microwave background.

Are you on the edge of your seat yet? What did *BOOMERANG* find, you ask? It turns out that the scale size of the detected fluctuations was about 1 degree.

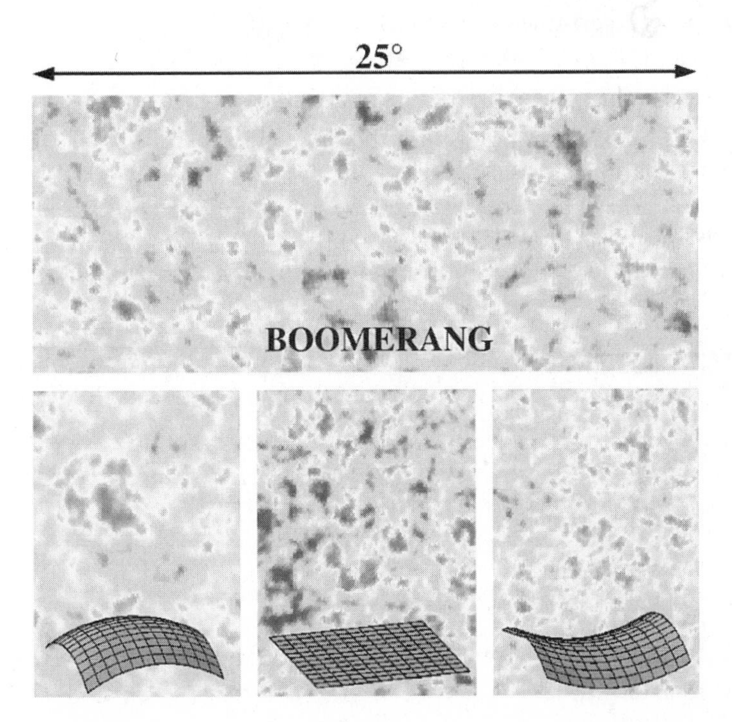

The scale size of the temperature fluctuations in the microwave background can tell us about the geometry of the universe. A flat universe should have fluctuations with scale sizes of about 1 degree. In a closed universe the fluctuations would be larger, and in an open universe, smaller. The BOOMERANG *data (center image) strongly suggested that we live in a universe with a flat geometry.*

(Image from NASA)

In a flat universe, one that has critical density, or $\Omega_o = 1$, the scale of the fluctuations in the cosmic microwave background (CMB) is expected to be about 1 degree, or twice the size of the full moon. Fluctuations smaller or larger than 1 degree mean that the universe has a curvature. In a closed universe the fluctuations would be larger than 1 degree, and in an open universe, they would be smaller. The *BOOMERANG* result implied a flat universe.

Fasten Your Seatbelts, We're Accelerating

While the *BOOMERANG* scientists were lending credence to the notion of a flat universe, other astronomers were busy observing very distant sources and making some surprising discoveries of their own concerning not the shape of the universe, but its age and its fate.

Many cosmologists have been intensely interested in the correct value for the Hubble constant. This constant is basically the "slope" of the line describing how fast galaxies are receding from one another as a function of their separation. Recent measurements of high redshift supernovae have fixed the slope of this line to be 71 km/sec/Mpc. That is, for every million parsecs of distance from us, galaxies are moving away from us at 71 km/s. Astronomers determined this number from careful measurements of the distances to objects and the speed that they are moving away from us.

The reason for the interest is that the slope of this line can be used to tell us the age of the universe. And everybody wants to know the age of the universe, right?

But there is a more subtle effect as well. Ever since the Big Bang was proposed, astronomers have been curious about the fate of the universe. Will the universe expand forever, or will it eventually halt its expansion and collapse into another fireball?

One way to answer this question is to compare the expansion rate of the universe now (using galaxies close to us) to the expansion rate long ago (using galaxies far away from us). With good data, we should be able to figure out if the expansion of the universe is slowing down or not. And if it's slowing down at the right rate, then we know that we are in for a Big Crunch sometime in the future.

Close Encounter

Some recent discoveries have a bearing on the question of the eventual fate of the universe. Two groups of astronomers (one at the Harvard-Smithsonian Center for Astrophysics, and the other at Berkeley Labs) using the Hubble Space Telescope have found evidence that the universe is likely to expand forever—that the universe doesn't have enough mass in it to cause the expansion to halt.

They have used what we call Type Ia supernovae (cataclysmic explosions that result when a white dwarf star borrows a bit too much material from its binary companion) as standard candles. Type Ia supernovae are so incredibly luminous that they enable us to see farther than just about any other phenomenon we know of. Using these supernovae, astronomers have determined that the expansion of the universe (as described by Hubble's Law) doesn't seem to be slowing down at all. In fact, it appears to be speeding up.

What Type of Supernovae Would You Like?

Comparing these two rates of expansion is exactly what a pair of competing research groups—the Supernova Cosmology Project and the High-Z Supernova Search Team—started to do in the mid-1990s. Their goal was to find a relatively common astronomical object that had a known luminosity that they could use as a "standard candle" to great distances. That meant it had to be very bright, and Type Ia supernovae turned out to be the perfect objects.

Type Ia supernovae are the explosions of white dwarfs that have been stealing material from a binary companion. When the mass of the thieving star exceeds 1.4 solar masses, it will explode as a Type Ia supernova. These events are 10 to 100 times brighter than a Type II supernova, which is the collapse and explosion of a normal high-mass star.

Astronomers had studied Type Ia supernovae long enough to know they had a known "light curve," or brightening and fading as a function of time, and most importantly, nearby supernovae of this kind seemed to have a very regular peak luminosity. So if one could just identify Type Ia supernovae (which can be done by analyzing their optical spectra), then he could use their apparent brightness to determine the distance to the galaxy in which the supernova was detected. The beauty is that a Type Ia supernova can be seen up to billions of light-years away.

This Can't Be Right

The two teams studied the data from a total of about 50 high-redshift Type Ia supernovae and were surprised to find that the expansion of the universe was not slowing down at all. What was truly shocking was that the expansion of the universe was, in fact, speeding up. All the evidence from the high-redshift supernovae, which enabled researchers to trace the Hubble expansion back to the earliest times in the universe, indicated that the universe was accelerating in its expansion, and it was expanding more rapidly now than it had in the past.

This was a result that no one had expected. Astronomers understood an expansion that was slowing down as indicating that the universe had critical density and that the force of gravity was winning out. But an *accelerating* expansion? What did *that* mean?

Blunder or Brilliance?

Remember how Einstein had to initially propose a "cosmological constant" lest his theoretical universe collapse under its own gravity? The cosmological constant kept his universe pumped up by providing a force that pushed things apart.

When Edwin Hubble discovered the recession of galaxies in the 1920s, however, the cosmological constant was put away, deemed unnecessary, and an abashed Einstein called it his "biggest blunder." Then in 1998, with the discovery of the accelerating expansion of the universe, scientists pulled the cosmological constant out of the drawer and dusted it off.

Could the force that it described (which followed naturally from Einstein's theory of general relativity) be the force that was pushing things apart?

Now recall that despite their best efforts, astronomers had at the time been unable to account for more than about 30 percent of the critical density of the universe. Including luminous and dark matter, there is about 70 percent of the critical density entirely unaccounted for. Yet the evidence from the cosmic microwave background seemed to indicate that the universe had critical density and a flat geometry.

It is possible that the other 70 percent of the density of the universe is not in its mass but in some form of energy. Recall that energy and mass are in some sense interchangeable, as Einstein famously described with the equation $E = mc^2$. In fact, the latest observations support about 25 percent dark matter and 75 percent dark energy, giving the universe its critical density.

Einstein's "blunder" might just point the way. If the expansion of the universe is accelerating and the universe has "critical density," these might both be explained by energy contained in the vacuum of space itself.

Why So Critical?

Why should we expect that the universe has critical density? Does some human aesthetic bias compel us to envision a universe that has the density required to balance gravity exactly?

Not really. Even before the *BOOMERANG* experiment found that the scale of the fluctuations in the CMB was about 1 degree—and therefore consistent with a flat universe and critical density—discussions occurred among cosmologists that went something like this: the value of Ω (the density in mass and energy) could be anything. It is very unlikely that the value would be so close to 1 (remember, observations of the universe indicate that dark and luminous matter can account for about 30 percent of critical density), and not actually be 1.

But with *BOOMERANG* there was clear evidence that the universe had critical density. And more evidence has been coming in.

This Just In: The Whole Sky

The *BOOMERANG* mission did in some sense scoop the NASA *MAP* (Microwave Aniotropy Probe), now called the *WMAP* (Wilkinson Microwave Anisotropy Probe). *BOOMERANG* made the first determination of the scale of the fluctuations in the CMB, but the high-resolution image of the fluctuations in the CMB of the whole sky had to wait for the *WMAP*, which was launched in June 2001.

In February 2003, NASA and the *WMAP* team released the first high-resolution, all-sky images of the cosmic microwave background. These results have provided stunning confirmation of the *BOOMERANG* result and have placed even tighter constraints on the history of the early universe. Another important result from the *WMAP* is that the first stars appear to have formed when the universe was only about 200 million years old, much earlier than we had previously thought.

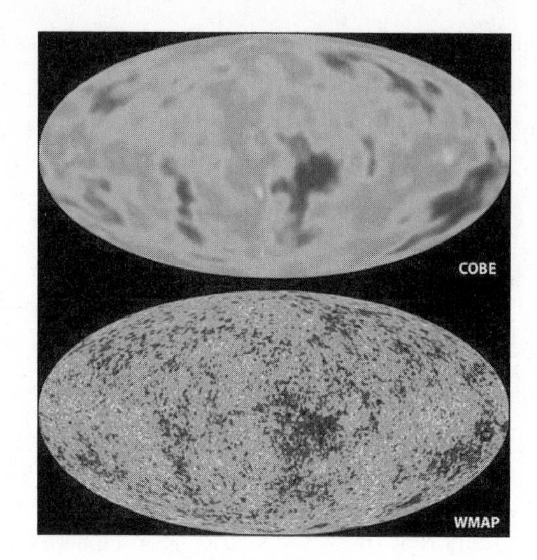

*The Wilkinson Microwave Anisotropy Probe (*WMAP*) has imaged the entire sky with less than 1 degree resolution, similar to the* BOOMERANG *image. This image shows the comparison between the* COBE *(top) and the* WMAP *(bottom) results.*

(Image from WMAP/*NASA)*

The Future's a *SNAP*

The high-redshift supernovae observations that began with ground-based observations in the 1990s have moved in part to observations with the Hubble Space Telescope. In the future, research teams hope that a dedicated satellite instrument still in its developmental stages, called *SNAP*, or the SuperNova/Acceleration Probe, will carry out the observations.

One of the limitations of the current high-redshift supernovae work is that it is often difficult to get time on telescopes for follow-up observations on short notice. Follow-up observations are required to determine the type of a detected supernova and to track the evolution of the light curve. As a dedicated instrument, *SNAP* would resolve many of these observational issues and add to our list many more supernovae at very high redshifts.

Putting It All Together

Let's now pull together all the issues that scientists have clarified about the early history of the universe in the past five years, a half decade that has been remarkable for astronomy:

- Based on the *WMAP* results and other recent observations, the universe has an age of 13.7 billion years, plus or minus only about 100 million years.
- The first stars in the universe started fusing hydrogen only 200 million years after the Big Bang, much earlier than previously thought.

- The universe consists of 4 percent "normal" atoms (the stuff we are made of), 23 percent dark matter, and 73 percent dark energy.
- The expansion rate of the universe gives us a value for the Hubble constant (H_0) of 71 km/sec/Mpc.

In 1920, Robert Frost wrote a poem called "Fire and Ice." Frost began, "Some say the world will end in fire, Some say in ice." We might now have a definitive answer.

In the end, it looks as if we are just a tiny part of a universe that is going to expand forever, at an ever-increasing rate. The universe, then, will end in ice. Of course, that still leaves us with absolutely no idea what 96 percent of the universe is—right now, while it's still in business. In the final chapter, we wonder if anyone else out there in the universe is pondering its content and its fate. And if so, will we ever be able to get together and talk about it?

The Least You Need to Know

- The cosmic microwave background (CMB), which is our insight into the earliest time of the universe, indicates that the universe is not perfectly homogeneous; tiny fluctuations in density were present even when the universe was in its infancy, only 300,000 years old.
- The scale of the fluctuations in the CMB is about 1 degree: the size of fluctuations expected in a "flat" universe.
- The expansion of the universe appears to be accelerating.
- Combining the results of high-redshift supernovae and the fluctuations in the cosmic microwave background suggests that most of the density of the universe is not in visible matter or dark matter, but in dark energy, whatever that might be.

Chapter 18

Where Is Everybody?

In This Chapter

- The Fermi Paradox
- Definitions of life
- Earth: unique or typical in its ability to sustain life?
- Other life in our solar system
- The case for life on Mars
- Life in our Galaxy: the Drake Equation

Enrico Fermi, the Italian-born physicist whose research team created the first controlled nuclear chain reaction, posed a question in the 1950s that has since become known as the Fermi Paradox: if life is abundant in the universe and the universe is both vast and old, where is everybody?

The question goes to the heart of something that humans have wondered about for a long time. Why do we feel so alone?

We might consider a number of questions to resolve the paradox. Is the universe perhaps teeming with life that can't get to us because of the limitations and cost of space travel? Are we the first wave of intelligent life in the Milky Way? Or are others so advanced as to be simply uninterested in us? This chapter confronts some of these questions and may help you decide your own answers to them.

In the past four centuries, we have learned that nothing is special about our position in the cosmos. We inhabit the third planet orbiting an average star in a solar system tens of thousands of light-years from the center of a normal spiral galaxy. And a few hundred billion other stars are also in the Galaxy. Granted, not all of them could have solar systems like our own, but the history of our understanding of our place in the universe should tell us that an assumption of mediocrity is probably a good idea.

If we're not special, why has no one contacted us? Do we smell bad? Is it our hair? Or is our planet truly the only repository of life in the universe? Let's take an inventory of what we know, and look at what it might tell us about our place in the universe as intelligent, curious beings.

What Do You Mean by "Alone"?

Sitting in your home at night, watching television, with no other humans present, you might consider yourself alone. But are you really? There are microbes crawling on your skin; bacteria are in your intestines; and your cat is sprawled across your lap. Silverfish inhabit the damp cabinet underneath your kitchen sink, and yet you might still consider yourself "alone." Why?

Although other living things are close to you, some even on or in your own body, you cannot communicate with them. Your cat may come close to communicating, but even she has limitations beyond appreciating being fed and scratched.

Now reconsider the larger question. Is the mere existence of other life in the universe enough for us not to feel alone? Would there be any comfort in knowing that the universe is teeming with bacterial life? Or will our cosmic loneliness be ended only by the presence of other intelligent life, life with which we can potentially communicate?

... If You Call This Living

So in thinking about the prospects for life beyond our planet, most of us probably contemplate two distinct subjects: intelligent life—life capable of creating civilizations—and, well, just plain life.

To start with, just how plain can life get and still be called life?

You know how a shrink-wrapped box of software has "minimum requirements" printed on its side? And it tells you how many megabytes of RAM you need, how much hard drive space, and so on? Well, life seems to have some minimum requirements as well.

Most biologists agree that the requirements for something to be classified as a living organism include …

- The ability to reproduce
- The ability to obtain nourishment or energy from the environment
- The ability to evolve or change as the environment changes

The late, great evolutionary biologist Steven J. Gould called the current age on Earth the "Age of Bacteria," and he made the case that, in terms of diversity and adaptability, "bacteria rule." The science fiction writer Kurt Vonnegut Jr. once suggested that the purpose of human beings is to bring viruses to other planets. Maybe he was joking. Maybe.

Bacteria and viruses might be plentiful, but what about more macroscopic beings? Could these be found—or even be common—on other planets?

Do You Like Your Earth Served Rare?

The previously listed requirements are minimal and, for example, say nothing about movement, specific senses, or even consciousness—attributes that narrow and refine the definition of life. We know that life on Earth ranges from tiny viruses—which hardly seem alive at all when inactive and which most biologists do not classify as living—to simple one-celled organisms, to the whole range of more complex plant and animal life, to ourselves.

Life at its most basic seems to require very little: the presence of a few chemical elements, the absence of certain harmful substances, and the availability of tolerable environmental conditions. Some scientists believe the existence of these basic requirements on Earth is a sort of cosmic fluke—a rare, possibly even unique, lottery jackpot win. But most assume that Earth, although special to us, is—within the Galaxy—mediocre. The belief that nothing is uncommon, let alone unique, about the conditions on Earth that support life is sometimes called the *assumption of mediocrity*.

As far as we can tell, Earth does not possess any special elements or conditions that aren't easily available elsewhere in the Galaxy and the universe. Not only, then, is there nothing to have prevented life from evolving elsewhere in the universe, but with hundreds of billions of stars in our Galaxy alone, we have every reason to believe that life *has* indeed sprung forth elsewhere.

There are even recent suggestions that life might have originated on another planet of our solar system—namely Mars—and was brought to Earth on the backs of asteroids. Other researchers have been able to generate structures that could be the precursors to cell walls in conditions that simulate the harsh environments of interstellar space. Life, once thought to be fragile and unique, appears to be quite a tough survivor.

Okay, fine. But even given all of these hopeful signs for life in the universe, Fermi's question remains unanswered: "Where *is* everybody?"

The Chemistry of Life

According to those who study high-redshift supernovae, the universe is about 14 billion years old. Earth, like other planetary bodies in the solar system, appears to be about 4.5 billion years old, and the fossil record shows that Earth was not devoid of life for very long, perhaps for only a few hundred million years after it coalesced from the solar nebula.

The oldest fossils known on Earth, Precambrian microfossils, are from western Australia and are about 3.5 billion years old. They resemble currently living anaerobic bacteria, bacteria which do not require free atmospheric oxygen. The first fossils indicating the presence of life requiring free oxygen are about 2 billion years old, which means that Earth's atmosphere underwent a huge change in the intervening billion and a half years. The fossil record tells us that life arose relatively quickly on the surface of Earth and went through waves of diversity and evolution on timescales of tens of millions of years.

Astro Byte

Volcanic activity on the early Earth sent many gases coursing into the atmosphere. Its early carbon dioxide atmosphere literally arose from within. Earth had sufficient mass to hold onto the heavier molecules in its atmosphere but not the lighter hydrogen and helium. Earth was a warmer place in the past, warmed by its greenhouse gases.

One glaring mystery remains: how did life get a foothold in the first place?

Most scientists who study life on early Earth propose that sometime before the first life forms arose, during Earth's initial several hundred million years, the simple molecules present (nitrogen, oxygen, carbon dioxide gas, ammonia, and methane) somehow formed into the more complex amino acids that are the chemical building blocks of life. The turbulent youth of Earth provided the energy that caused the transformation of simple elements and compounds into the building blocks of life. Also remember that, in its infancy, Earth was orbited much more closely by the Moon, so that after liquid water was present on its surface, the tides would have been enormous—perhaps 1,000 feet up and down as the planet rotated once every 6 (not 24) hours. This vigorous mixing of the primordial soup (shaken, not stirred) might have had an important effect on the origin of life, providing a "workbench" on which somewhat random chemical experiments could occur.

After amino acids and nucleotide bases were available (and observational evidence of star-forming regions suggests that highly complex molecules are present in interstellar space), the next step up in complexity would have been the synthesis of proteins and genetic material.

The DNA molecule is made up of nucleotide bases called genes. Strung together, the genes tell our cells what to do when they reproduce and make us different (in some ways) from, say, earthworms. The DNA molecule is the most durable, portable, flexible, and compact information storage device we know of, and we are just beginning to unravel its mysteries.

But this is not to say that DNA is the only way life could propagate and could have propagated. Perhaps there are myriad other means by which life can encode ways to make more of itself. Nevertheless, a starting point is to understand the genetic material that supports life on Earth, and we are still in the early phases of that understanding.

Although plenty of planets might be similar to Earth, to estimate the likelihood that molecules on any given planet will combine to form amino acids and nucleotides, let alone proteins and DNA, is extremely difficult. *Astrobiologists* have wondered: Are there fundamental biological laws in the same way that there are fundamental physical laws? Do the basic, universal rules of chemical equations inevitably lead to information storage systems like DNA? Or is DNA a fluke peculiar to our planet? This is where the questions, and the debates, begin.

def•i•ni•tion

Astrobiologists are scientists who contemplate questions (like the origin and evolution of life on Earth) that arise at the intersection of these two fields. Astrobiology is a relatively new discipline, combining the expertise of astronomers and biologists.

Close Encounter

In 1953, scientists Stanley Miller and Harold Urey decided to see if they could duplicate, experimentally, the chemical and atmospheric conditions that produced life on Earth. In a 5-liter flask, they replicated what is thought to be Earth's primordial atmosphere: methane, carbon dioxide, ammonia, and water. They wired the flask for an electric discharge, the spark intended to simulate a source of ultraviolet photons (lightning) or other form of energy (such as a meteor impact shock wave, which must have been common in the early solar system). The Miller-Urey Experiment didn't produce life, but it did create a collection of amino acids (building blocks of proteins), sugars, and other organic compounds. Thus the Miller-Urey Experiment implied an extension of the assumption of mediocrity. Not only could we reasonably conclude there must be similar planets in the Galaxy but also that they must have the chemical elements necessary for life, and random energy input alone could trigger the synthesis of the building blocks of life.

All life on Earth is carbon-based. That is, its constituent chemical compounds are built on combinations of carbon atoms and, furthermore, developed in a liquid water environment. Even creatures such as ourselves, who do not live in water, consist mostly of water, and the presence of liquid water is vital to our continued existence. If life on Earth developed first in the oceans, we, billions of years later, individually develop as fetuses in a tiny oceanlike womb; then we—being mostly water—carry these oceans within us through our lives.

Life on Mars

Science fiction has long portrayed Mars as home to intelligent life, and the American astronomer Percival Lowell, early in the twentieth century, created a great stir with his theory of Martian "canals." Actor-director-writer Orson Welles triggered a nationwide panic with his 1938 radio dramatization of H. G. Wells's *War of the Worlds*, about a Martian invasion of Earth. People of Earth seemed primed to believe that life might exist on Mars.

Given our fascination with the red planet and its proximity, it is no wonder Mars has been the target of a number of unmanned probes. In the mid- to late 1970s, the *Viking* probes performed robotic experiments on the Martian surface, including tests designed to detect the presence of simple life forms such as microbes. These tests yielded apparently positive results, which, however, were subsequently reinterpreted as false positives resulting from chemical reactions with the Martian soil. Then, on August 7, 1996, scientists announced that, based on its chemical composition, a meteorite recovered in Antarctica had originated on Mars and possibly contained evidence of fossilized bacterial life.

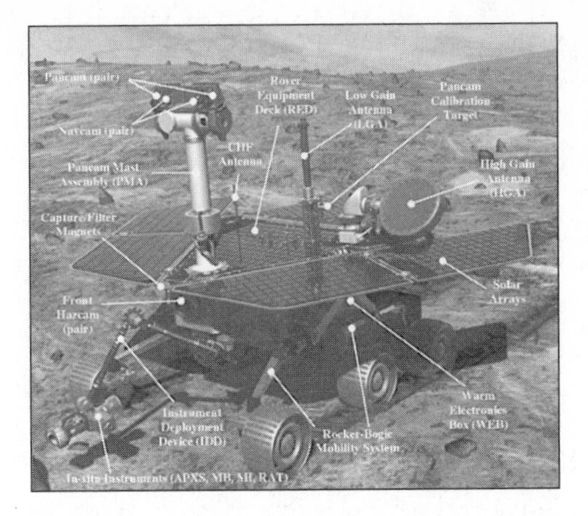

The two Mars Exploration Rovers have been functioning as robotic scientists on the surface of Mars, extending the reach of human understanding, and far outlasting their expected lifetimes.

(Image from NASA)

The *Mars Pathfinder* (1997) and *Mars Global Surveyor* (1998) missions have taken detailed panoramic views from the Martian surface and high-resolution satellite images of the surface. Both of these missions have found evidence of the presence of liquid water on the Martian surface at some time in the past. And in January 2004, two Martian rovers NASA launched in the summer of 2003 touched down on the surface of Mars and found geological evidence of the presence of water on the planet in the past.

The rover Spirit, *almost three years after launch, recently uncovered evidence of an ancient watery environment in Gusev Crater: fine-grained silica (SiO_2) exposed by dragging one of its wheels. Silica sediments are found on ancient Earth, deposited in coastal waters.*

(Image from NASA/JPL-Caltech/Cornell)

All of these results are intriguing, but so far, none of the Martian probes has found incontrovertible evidence of life (past or present) on Mars. In the past, the Martian atmosphere, now very thin, was likely thicker and, as a consequence, the planet's surface was warmer and wetter. So, possibly, microbial life once existed on Mars. Under the planet's current cold, dry, and generally harsh conditions, however, the presence of life today is highly unlikely.

Hello! Is Anybody Out There?

So life beyond Earth but within the solar system seems unlikely. But that's okay. There's still the rest of the Galaxy, right?

With hundreds of billions of stars in our own Galaxy, odds are there must be many other planetary systems. Observations of the "wobble" of stars indicates the presence of well over 100 stars with planets around them. Recent research has shown that stars with high metallicity (determined by the relative abundance of iron in the star) are much more likely to have planets. But that narrows the odds only slightly. And if one of these planets orbits at a constant distance from its host star at a distance at which liquid water could exist, couldn't life exist there as well? How, exactly, do we determine the odds?

An Equation You'll Like

Given the billions of stars in our Galaxy, the existence of life somewhere in the Milky Way is pretty likely, but do we have any way to get a rough idea of just how likely? Not every star in the Galaxy is a good prospect for life. Some stars are so massive they don't last long enough, and others are not hot enough to warm any planets that might be close by. Others formed long ago when there were not as many metals present, and others are in binary systems in which (simulations show) planets have a harder time surviving.

Perhaps 15 percent of the stars in the Galaxy have the proper mass to be just luminous enough—but not too luminous—to support habitable planets. In 1961, an astronomer named Frank Drake attempted to quantify the odds that intelligent life capable of communication exists elsewhere in the Galaxy and tried to get at least a rough answer to part of the Fermi Paradox.

Drake proposed an equation to estimate the possible number of civilizations in the Milky Way. The Drake Equation contains a number of highly uncertain terms—variables that must simply be guessed at; nevertheless, it is a useful way of breaking down a complex question into smaller, better-differentiated blocks, which may become more certain as technology and understanding improve. Some of the variables that were highly uncertain in 1961—the number of stars with planets, for example—have become more certain as further relevant observations were made.

Here's what the Drake Equation looks like:

$$N = R_* \times f_p \times N_p \times f_e \times f_l \times f_i \times f_c \times L$$

And here's what the terms of the Drake Equation stand for:

- N (the left-hand side of the equation) is the number of civilizations in our Galaxy with which we should be able to communicate via some sort of signal.
- R_* is the rate at which our Galaxy forms solar-mass stars—its productivity in solar masses per year.
- f_p is the fraction of stars with planetary systems.
- N_p is the average number of planets per star.
- f_e is the fraction of Earthlike planets (planets suitable for life).
- f_l is the fraction of these planets on which life actually develops.
- f_i is the fraction of these planets on which intelligent civilizations arise.
- f_c is the fraction of these planets on which technological civilizations develop.
- L is the lifetime of a civilization in years.

A Careful Look at the Equation

Perhaps the most important term in the equation is L, the lifetime of a civilization. If civilizations last for many millions of years, then our own Galaxy could be teeming not only with life but also with advanced civilizations. As Fermi and others realized, the universe has been around for a long time. The types of stars around which civilizations could evolve have been present since relatively soon after the Big Bang, and our planet took about 5 billion years to produce us. But Earth didn't form until the universe was almost 10 billion years old. If the likelihood of all of these variables is high, intelligent civilizations should have colonized the Galaxy by now. So where are they?

Here's one somewhat depressing possibility: if *L* is rather small (even hundreds or thousands of years), then the Galaxy might be a lonely place indeed. Civilizations, like candles, might be lit frequently, but if they do not burn for very long, only a few others (or no others) will be burning simultaneously with us.

Let's look at some of the other key terms in the equation.

Galaxy Productivity

Astronomers figure that a minimum of 100 billion stars populate the Milky Way. We can estimate that about 10 percent of these stars—or about 10 billion stars—are solar-mass stars that are not in binary systems. If the Galaxy is about 10 billion years old, then the star formation rate is about 10 billion stars divided by 10 billion years, or 1 solar-mass star each year.

Do They All Have Planets?

When the Drake Equation was first written down, the only planetary system we knew about was our own. But astronomers have expended great effort in the past three decades to uncover other planetary systems. As noted previously, we now know many other planetary systems (over 200 at last count) exist, and a recent study based on statistical corrections to planetary searches estimates that at least 25 percent of solar-mass stars have planetary systems.

Welcome to the Habitable Zone

Estimates of the typical number of planets per planetary system, based on numerical modeling, range from 5 to 20. The number is highly dependent on the simulation that is run, but that shouldn't be surprising, since the formation of a planetary system has some degree of randomness.

Our own solar system, of course, has eight planets and a host of "dwarf planets," including Pluto, which are now being discovered. We have observations of other planetary systems, but we must remember that, at present, we can detect only the most massive planets orbiting these stars. We have no counts of the number of smaller planets in these systems.

However, many planets do form around a star; astronomers call a doughnut-shaped region around all stars the *habitable zone*. This zone is the space between some inner and outer distance from the star where a star with a given luminosity (rate of energy production) produces a temperature that's "just right."

Just right for what? Well, just right for liquid water to exist without freezing or boiling away. We will estimate that 10 percent of the planets around a star will fall within the habitable zone. Recent work has narrowed this zone by introducing the concept of a "continuously habitable zone." This is the distance from a star (throughout its changeable lifetime) at which liquid water could exist. It is interesting to note that although Venus, Earth, and Mars fall in the habitable zone of the Sun, only Earth falls in the narrower *continuously* habitable zone.

Astronomer's Notebook

Three planets are in the habitable zone of the Sun—Venus, Earth, and Mars. The habitable zone is the distance from the star at which liquid water could survive. So apparently being in the "zone" is not sufficient to support life!

Primordial Soup du Jour

Although astronomers such as Carl Sagan (1934–1997) have assumed that life develops on virtually every planet capable of supporting it, estimates of this probability vary widely. Until scientists produce a model whereby basic elements can self-organize into reproducing "proto-life," the odds of life arising given the right conditions are anybody's guess.

The fraction of habitable planets on which life arises, f_l, is expressed as a decimal, with the number 1 indicating that there is a 100 percent chance that life will develop on a habitable planet. Estimates range from Sagan's optimistic 1 down to 0.000001—that is, perhaps only 1 out of every 1 million habitable planets actually develops life. This number is highly uncertain. If robotic probes in the solar system were to find that microbial life once existed on Mars, then this might cause us to raise the odds that life will arise given the right conditions.

As we will see later, the probability that simple life arises might be high, but might only be undone later in the equation by the difficulties presented to long-term survival of complex life.

You Said Intelligent Life? Where?

Beyond this point in the equation, the terms become very uncertain. Some scientists believe that it is possible that biological evolution as observed on Earth is a universal phenomenon; given enough time, life anywhere it exists will tend to evolve toward greater and greater complexity.

If this is the case, then we might assume that, given enough time, intelligence will emerge wherever there is life. Others, however, believe that intelligence is not indispensable to survival and that, therefore, it is not the inevitable byproduct of natural selection and evolution. Are bacteria intelligent? Not particularly, and they've done pretty well for themselves here on Earth.

Turn on the Radio

For the fraction of intelligent life forms that develop civilization and technology (f_c), estimates vary widely. How could they not? We have now entered the realm of psychology and social science, disciplines that have a hard enough time understanding humans, much less utterly conjectural life forms.

Some argue that given time, technology and civilization are inevitable developments of intelligence. Others make just as compelling a case that the connection is not inevitable and, consequently, put f_c much lower. Other complicating factors involve the correlation between intelligence and aggression. Are intelligent life forms destined to destroy themselves? Or do a few select civilizations enter a time of peace, a *pax planetaria?*

Finally, we cannot assume that all civilizations will have the *desire* to communicate. Maybe civilizations develop virtual realities much more interesting than the real universe and develop beyond physical, biological bodies altogether. This term in the Drake Equation has been rich territory for mining by good science fiction writers.

The End of the World As We Know It

The final term is perhaps both the most interesting and the most uncertain of them all: how long do civilizations capable of communication endure? To estimate how many civilizations there are in the Galaxy at any given time, we must know how long they last. Once lit, how long does the candle of civilization burn, at least on average? Consider our own civilization, one with a recorded history of no more than 5,500 years and one that has been highly technological (possessing the atomic bomb and radio telescopes) for less than a century.

Planets and stars have long lifespans, but we have only a little over 5,000 years of documented experience with civilization. What is most disquieting is that we have reached a point of technological development that enables us, at the touch of a few buttons, to destroy our civilization in a thermonuclear fireball here and now. And the solar system is still a place where things are out of our control. Were an asteroid like the one that may have struck Earth about 65 million years ago to hit us again, we would fare no better than the dinosaurs. Fifty million years after that impact, highly evolved ants might be excavating our fossilized bones and speculating on the cause of our demise.

So what is the average life span of a civilization? A thousand, five thousand, ten thousand, a million years? Even if, over time, there have been many, many civilizations in our Galaxy, how many are there now?

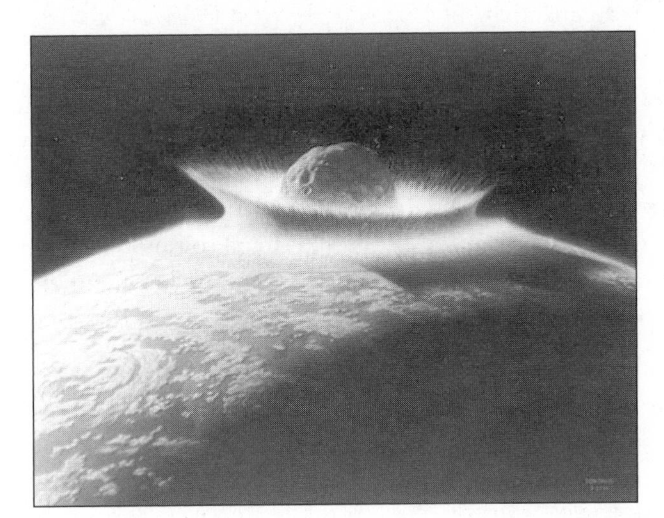

Were an asteroid like the one that may have doomed the dinosaurs to hit Earth tomorrow, we would fare no better than they did. Civilizations face internal and external challenges to their survival.

(Image from NASA)

Close Encounter

If we approach any variable in the Drake Equation pessimistically, assigning it a low value, we end up with the possibility that there are very few technological civilizations in the Galaxy. But if we plug into the equation the most optimistic estimates for all the variables (such as life will always develop; intelligence will always evolve; technology will always arise), we end up finding that the number of technological civilizations in the Galaxy is equal to the average lifetime of such a civilization.

If the average technological civilization survives 10,000 years, we can expect to find 10,000 civilizations in the Galaxy at any time. If 5,000 years is the norm, 5,000 civilizations would be expected. If a civilization usually lasts millions of years, then, assuming high values throughout the rest of the equation, we might expect to find a million civilizations in the Milky Way. But even in this optimistic scenario, the distances between civilizations remain enormous.

Traveling to the location of another civilization in the Galaxy is a highly impractical proposition. Even in a best-case scenario, civilizations in the Galaxy must be hundreds of light-years apart, and, using current technology, getting to the nearest star would take us tens of thousands of years. A civilization might be long gone by the time we got there—to say nothing of having to cope with the mother of all jet lags at the end of the trip.

Wouldn't it be tragic to travel to another star system to find that they had done themselves in during the thousands of years it took us to get there?

What We Look For

Given the vast distances that must separate civilizations even in the most optimistic scenarios, it is unlikely that we will be traveling to distant civilizations anytime soon. But the urge to detect some sign of other intelligent life in the universe persists, and since the 1960s, dedicated groups of researchers have trained radio telescopes skyward in the hope of receiving broadcasts from extraterrestrial sources. Searches have varied from targeted (directed toward other Sunlike stars) to random (piggybacking on other astronomical observations). So far, none of the searches have detected a repeating signal of extraterrestrial origin.

Astro Byte

Why use radio signals? Why not use an optical laser pulse sent out into space? It is certainly possible, but because the dust in the plane of our Galaxy absorbs optical photons very well, the energy requirements for optical communication would be much greater. Radio waves, on the other hand, pass through the plane of the Galaxy as if it weren't there. We (and our potential neighbors) can cast a cheaper, wider net in radio waves.

"Earlier on *Survivor* ..."

The project of monitoring the heavens for radio broadcasts seems to assume that a technological civilization would have made the decision intentionally to broadcast a repeating signal. Given that we Earthlings are doing no such thing, this assumption seems a bit shaky.

There is no need to assume that the civilization actually *wants* to communicate and is intentionally broadcasting to other worlds. After all, we on Earth have been broadcasting radio waves since the early twentieth century and have been doing so intensively for about 70 years.

Although radio signals at longer wavelengths do not penetrate beyond our atmosphere, those in the higher frequencies, FM radio and television, are emitted into space. We don't *intend* this to happen, but it is an inevitable by-product of our technological civilization—at least until everything comes through optical fibers. Perhaps some other civilization is producing a similar by-product. Our first glimpse of another civilization might be the equivalent of its sitcoms, advertising, reality TV, and televangelists. And yet we keep on listening.

The SETI Search

Monitoring the heavens for artificial extraterrestrial broadcasts is a daunting task—akin to looking for a needle in a haystack. Where should we look? What frequencies should we monitor? How strong will the signal be? Will it be continuous or intermittent? Will it drift? Will it change frequency? Will it even be recognizable to us?

To say the least, radio frequency searches are time- and equipment-intensive. Frank Drake conducted the first search in 1960, using the 85-foot antenna at the National Radio Astronomy Observatory in Green Bank, West Virginia. He called his endeavor Project Ozma, after the queen of the land of Oz. This search ultimately developed into Project SETI (Search for Extra-Terrestrial Intelligence), which, despite losing federal funding in the 1990s, remains the most important sponsor of search and research efforts.

The SETI Institute is a private, nonprofit group based in Mountainview, California. When NASA-based SETI funding was cut in 1993, SETI consolidated much of its effort into Project Phoenix (risen, like the mythical bird, from the ashes of the funding cut), a program that began in 1995 and monitors 28 million channels simultaneously. SETI hopes eventually to monitor 2 billion channels for about 1,000 nearby stars. Computer software alerts astronomers to any unusual, repeating signals. So far, no repeating signals have been discovered.

Down at the Old Water Hole

SETI researchers also guess that civilizations might choose intentionally to broadcast their presence to their neighbors, sending some sort of radio-frequency beacon into space. What portion of the spectrum might they choose?

Close Encounter

Want to do your bit for the Search for Extra-Terrestrial Intelligence (SETI)? Download a screensaver (at www.seti-inst.edu/science/setiathome.html) that might help to find that first ET. The program uses cycles on your computer (when not in use) to process one of the largest datasets mankind has ever assembled.

And if you just want to check the current status of SETI, go to their informative website: www.seti-inst.edu. The SETI Institute received a large financial boost from Microsoft co-founders Paul Allen and Nathan Myhrvold, who gave $11.5 million and $1 million respectively for the construction of a radio telescope array dedicated to the SETI search. Up to this point, searches have depended on cadging time on existing telescopes. The Allen Telescope Array (ATA) will be dedicated to the SETI project around the clock. The first stage of the ATA is under construction. By July 2006, the first 10 antennas of the ATA-42 (first stage) had been equipped with motors and receivers. When completed, the array will consist of 350 6.1-m telescopes.

In February 2007, the first astronomical images made with the ATA were released. In addition to being a SETI workhorse, the ATA will be a powerful astronomical radio telescope.

Researchers chiefly monitor a small portion of the radio wavelengths between 18 cm and 21 cm, called the *water hole.* The rationale for monitoring this slice of the radio spectrum is twofold. First, the most basic substance in the universe, hydrogen, radiates at a wavelength of 21 cm. Hydroxyl, the simple molecular combination of hydrogen and oxygen, radiates at 18 cm. If we combine hydrogen and hydroxyl, we get water.

If the symbolism of the water hole is not sufficiently persuasive to prompt extraterrestrial broadcasters to use these wavelengths, there is also the likelihood that this slice of the spectrum will be recognized as inviting on a more practical level. It is an especially quiet part of the radio spectrum. There is little interference here, a very low level of galactic "background" noise. Researchers reason that, at the very least, intentional broadcasters would see the water hole (which looks the same no matter where in the Galaxy one is located) as a most opportune broadcast channel.

Do We Really Want to Do This?

Hollywood has done a pretty fair job of reflecting the range of opinion on the psychological makeup of extraterrestrial beings. The 1950s and 1960s saw a number of movies about malevolent alien invaders, but then Steven Spielberg's *Close Encounters of the Third Kind* (1977) and *E.T.—the Extraterrestrial* (1982) suggested that contact with emissaries from other worlds might be thrilling, beneficial, and even heartwarming, not to mention boffo box office.

Astro Byte

The *Voyager* spacecraft each have onboard a golden plaque and phonograph record. Among other things, the plaque shows a man and woman, the trajectory of the spacecraft in the solar system, and a representation of the hydrogen atom. The phonograph record contains recordings of human voices sending a greeting in many languages and music. It is a message in a bottle cast into a vast ocean.

The close of the 1990s pitted *Independence Day*—in which aliens attempt to take over Earth—against *Contact*, in which communion with extraterrestrials is portrayed as a profoundly spiritual (if rather weepy) experience. And then *X-Files: Fight the Future* proposed that our entire planet is being prepared for massive colonization by extraterrestrials. In fact, the whole *X-Files* television series suggested a novel and surprisingly popular solution to the Fermi Paradox. Where is everybody? Hey, they're already here!

The point is this: not everyone is persuaded that reaching out is such a great idea. After all, even the most sociable among us lock our houses at night.

But, some argue, the point is moot; the radio cat has been out of the bag for more than six decades. An ever-expanding sphere of our radio broadcasts is moving out at the speed of light in all directions from Earth. The earliest television broadcast (which, unfortunately, includes images of Adolf Hitler at the 1936 Olympics) is now over 70 light-years from Earth. It's a bit late to start feeling shy.

On November 16, 1974, the giant radio dish of the Arecibo Observatory was used as a transmitter to broadcast a binary-coded message containing a compact treasure trove of information about humans. Reassembled, the message generates an image that shows the numbers 1 through 10; the atomic numbers for hydrogen, carbon, nitrogen, oxygen, and phosphorous; a representation of the double-helix of DNA; an iconic image of a human being; the population of the Earth; the position of Earth in the solar system; and a schematic representation of the Arecibo telescope itself. We have yet to receive a reply, but that should come as no surprise. Traveling at the speed of light, the message will take 25,000 years to reach its intended target, a globular cluster in the constellation Hercules. Maybe by then we will know how to greet our alien friends hospitably.

The Arecibo Radio Telescope in Puerto Rico broadcast the Arecibo Message in 1974. The message is 1,679 bits long, the product of two prime numbers, 23 and 73. When the zeros and ones are arranged in a rectangle with sides 23 and 73 units long, they make this picture, which shows among other things a human form, our number system, and the double helix of DNA.

(Image from NASA)

Some might view the Arecibo Message as the work of latter-day Quixotes jousting with cosmic windmills. Others might view it as an interstellar message cast adrift in a digital bottle, containing the very human hope of contact. Let's just hope that we haven't told them too much.

Coming Full Circle

We've come a long way since we first considered our astronomical ancestors looking up at the sky and wondering. We have come to the end of our story, a story that—as is obvious from daily newspaper, television, and Web stories—is still being written.

We live in an incredible age, when the evening news might report the discovery of accretion disks around black holes and distant Type Ia supernovae at the edge of the visible universe along with the doings of movie stars and musicians. Our book, we hope, will enable you to follow the news with interest, a critical eye, and curiosity about what is just beyond the horizon of our understanding.

The Least You Need to Know

- The chemicals and conditions on Earth that support life are probably not unique or even exceptional but common throughout the Galaxy and the universe.
- We can produce some of the basic building blocks of life in the laboratory and have even detected some of these basic blocks in interstellar space.
- Although the possibility of life beyond Earth within our solar system is very small, the existence of simple life elsewhere in our Galaxy is highly likely.
- The Drake Equation, a rough way to calculate in a reasoned manner the number of civilizations that might exist in the Milky Way, includes eight factors, some based in astronomy and well-determined, and others that are still highly speculative.
- Small but dedicated groups of researchers (including those of the SETI project) monitor the heavens for artificial radio signals of extraterrestrial origin, and the SETI Institute now has a dedicated telescope (the Allen Telescope Array) under construction.

Appendix A

Star Words Glossary

absolute magnitude *See* luminosity.

accelerating expansion Recent observations of very distant supernovae suggest that the expansion of the universe is accelerating; that is, the universe is expanding more quickly now than it did in the distant past.

accretion The gradual accumulation of mass; this usually refers to the build-up of larger masses from smaller ones through mutual gravitational attraction.

active galaxy A galaxy that has a more luminous nucleus than most galaxies.

altazimuth coordinates Altitude (angular distance above the horizon) and azimuth (compass direction expressed in angular measure from due north).

altitude *See* altazimuth coordinates.

angstrom Abbreviated A (or Å). This unit is used to measure very small size, equal to one ten billionth of a meter, or 10^{-10} meter.

angular momentum The rotating version of linear momentum. Depends on the mass distribution and rotational velocity of an object.

angular separation Distance between objects in the sky expressed as an angle (such as degrees) rather than in distance units (such as feet or meters).

angular size Size expressed as an angle (such as degrees) rather than in distance units (such as feet or meters).

annihilate Used as an intransitive verb by astronomers. Particles and antiparticles "annihilate" when they meet, converting their mass into energetic photons. Electrons and anti-electrons (positrons) were continually created and annihilated in the early universe.

anthropic principle Comes in different varieties, but the basic concept is that the universe must be the way it is (in terms of its fundamental parameters) or human beings would not be here to observe it.

apollo asteroids Asteroids with sufficiently eccentric orbits to cross paths with Earth (and other terrestrial planets).

apparent magnitude A value that depends on the distance to an object. *See also* luminosity.

arcminute One sixtieth of an angular degree.

arcsecond One sixtieth of an arcminute (1/3,600 of a degree).

assumption of mediocrity A scientific assumption that says we on Earth are not so special. The conditions that have enabled life to arise and evolve on Earth likely exist in many other places in the Galaxy and universe.

asterism An arbitrary grouping of stars, within or associated with a constellation, perceived to have a recognizable shape (such as the Teapot or Orion's Belt) that readily serves as a celestial landmark.

asteroid One of thousands of small, rocky members of the solar system that orbit the Sun. The largest asteroids are sometimes called minor planets.

astronomical unit (A.U.) A conventional unit of measurement equivalent to the average distance from Earth to the Sun (149,603,500 kilometers or 92,754,170 miles).

autumnal equinox The date (usually September 21) on which day and night are of equal length because the Sun's apparent course against the background stars (the ecliptic) intersects the celestial equator.

barred-spiral galaxy A spiral galaxy that has a linear feature, or bar of stars, running through the galaxy's center. The bar lies in the plane of the spiral galaxy's disk, and the spiral arms typically start at the end of the bar.

Big Bang The primordial explosion of a highly compact universe; the origin of the expansion of the universe.

binaries Also called binary stars. Two-star systems in which the stars orbit a common center of mass. The way the companion stars move can tell astronomers much about the individual stars, including their masses.

black body An idealized (theoretical) object that absorbs all radiation that falls on it and perfectly reemits all radiation it absorbs. The spectrum (or intensity of light as a function of wavelength) that such an object emits is an idealized mathematical onstruct called a black-body curve, which can serve as an index to measure the temperature of a real object. Some astronomical sources (such as stars) can be approximated as black bodies.

black hole A stellar-mass black hole is the end result of the core collapse of a high-mass (greater than 10 solar mass) star. It is an object from which no light can escape within a certain distance (*see* Schwarzschild radius). Although space behaves strangely very close to a black hole, at astronomical distances, the black hole's only effect is gravitational.

Bode's Law Also called the Titius-Bode Law, Bode's Law is a numerical trick that gives the approximate interval between some of the planetary orbits in our solar system.

brightness The measured intensity of radiation from an object. The brightness of astronomical objects falls off with the square of the distance.

brown dwarf A failed star, that is, a star in which the forces of heat and gravity reached equilibrium before the core temperature rose sufficiently to trigger nuclear fusion.

calderas Craters produced not by meteoroid impact but by volcanic activity. *See also* corona.

cardinal points The directions of due north, south, east, and west.

Cassini division A dark gap between rings A and B of Saturn. It is named for its discoverer, Gian Domenico Cassini (1625–1712), namesake of the mission currently exploring Saturn.

cataclysmic variable *See* variable star.

celestial equator This imaginary great circle divides the Northern and Southern Hemispheres of the celestial sphere.

celestial sphere An imaginary sphere surrounding Earth into which the stars are imagined to be fixed. For hundreds of years, people believed such a sphere (or bowl) really existed. Today, however, astronomers use the concept as a convenient fiction to describe the position of stars relative to one another.

chain reaction *See* nuclear fission.

circumpolar stars Stars near the celestial North Pole; from many locations on Earth, these stars never set.

closed universe A universe that is finite and without boundaries. A universe with density above the critical value is necessarily closed (*see* critical density). An open universe, in contrast, will expand forever because its density is insufficient to halt the expansion.

comet Also thought of as a "dirty snowball." This small celestial body, composed mainly of ice and dust, completes highly eccentric orbits around the Sun. As a comet approaches the Sun, some of its material is vaporized and ionized to create a gaseous head (coma) and two long tails (one made of dust, one of ions).

conjunction The apparent coming together of two celestial objects in the sky.

constellations These arbitrary formations of stars are perceived as figures or designs. There are 88 official constellations in the Northern and Southern Hemispheres.

convective motion A gas-flow pattern created by the rising movement of warm gases (or liquids) and the sinking movement of cooler gases (or liquids).

core The innermost region of a planet or star.

core hydrogen burning The principal nuclear fusion reaction process of a star. The hydrogen at the star's core is fused into helium, and the small amount of mass lost is used to produce enormous amounts of energy.

core-collapse supernova This extraordinarily energetic explosion results when the core of a high-mass star collapses under its own gravity.

core-halo galaxy *See* radio galaxy.

corona In astronomy, a corona might be a luminous ring appearing to surround a celestial body, the luminous envelope of ionized gas outside the Sun's chromosphere, or a large upswelling in the mantle of the surface of a planet or moon that takes the form of concentric fissures and that is an effect of volcanic activity. *See also* calderas.

cosmic microwave background (CMB) These highly redshifted photons left behind by the Big Bang are detectable today throughout all space as radio-wavelength radiation, indicating a black-body temperature for the universe of 2.73 K.

cosmological principle A cornerstone assumption about the nature of the universe. It holds that the universe exhibits two key properties: homogeneity (sameness of structure on the largest scale) and isotropy (appears the same in all directions).

cosmological redshift The lengthening of the wavelengths of electromagnetic radiation caused by the expansion of the universe.

cosmology The study of the origin, structure, and evolution of the universe.

crater This Latin word for "bowl" refers to the shape of depressions in the Moon or other celestial objects created (mostly) by meteoroid impacts.

critical density The density of matter in the universe that represents the division between a universe that expands infinitely (unbound, or open) and one that will ultimately collapse (bound, or closed). The density of the universe determines whether it will expand forever or end with a conflagration as dramatic as the Big Bang.

crust The surface layer of a planet.

dark energy This catch-all phrase is used to describe an energy of unknown origin that drives the apparent acceleration of the expansion of the universe.

dark halo The region surrounding the Milky Way and other galaxies that contains dark matter. The shape of the dark halo can be probed by examining in detail the effects its mass has on the rotation of the galaxy.

dark matter This catch-all phrase is used to describe an apparently abundant substance in the universe of unknown composition. Dark matter is 100 times more abundant than luminous matter on the largest scales.

differential rotation A property of anything that rotates and is not rigid. A spinning CD is a rigid rotator. A spinning piece of gelatin is not as rigid, and a spinning cloud of gas is even less so. For example, the atmospheres of the outer planets and of the Sun have equatorial regions that rotate at a different rate from the polar regions.

Drake Equation This equation proposes a number of terms that help us make a rough estimate of the number of civilizations in our Galaxy, the Milky Way.

dust lanes Dark areas sometimes visible within emission nebulae and galaxies; the term most frequently refers to interstellar absorption apparent in edge-on spiral galaxies.

dwarf planet Planets like Pluto in our solar system that orbit the host star, but unlike "normal" planets, are too small to clear the neighborhood of their orbit.

eccentric An ellipse (or elliptical orbit) is called eccentric when it is noncircular. An ellipse with an eccentricity of 0 is a circle, and an ellipse with an eccentricity of close to 1 would be very oblong.

eclipse An astronomical event in which one body passes in front of another so that the light from the occluded (shadowed) body is blocked. When the Sun, Moon, and Earth align, the Moon blocks the light of the Sun, resulting in a solar eclipse.

eclipsing binaries *See* visual binaries.

ecliptic The ecliptic traces the apparent path of the Sun against the background stars of the celestial sphere. This great circle is inclined at 23½ degrees relative to the celestial equator, which is the projection of the Earth's equator onto the celestial sphere.

electromagnetic radiation Energy in the form of rapidly fluctuating electric and magnetic fields and including visible light in addition to radio, infrared, ultraviolet, x-ray, and gamma-ray radiation. This energy often arises from moving charges in atoms and molecules, though high-energy radiation can arise in other processes.

electromagnetic spectrum The complete range of electromagnetic radiation, from radio waves to gamma waves and everything in between.

ellipse A "flattened" circle drawn around two foci instead of a single center point.

elliptical galaxy A galaxy with no discernible disk or bulge that looks like an oval or circle of stars on the sky. The true shapes of ellipticals vary from elongated ("footballs") to spherical ("baseballs") to flattened ("hamburger buns"). Elliptical galaxies consist of old stars and appear to have little or no gas in them.

emission lines Narrow regions of the spectrum where a particular substance is observed to emit its energy. These lines result from basic processes occurring on the smallest scales in an atom (such as electrons moving between energy levels).

emission nebulae Glowing clouds of hot, ionized interstellar gas located near a young, massive star. (Singular, *emission nebula.*)

ephemeredes These special almanacs give the daily positions of various celestial objects for periods of several years.

escape velocity The velocity necessary for an object to escape the gravitational pull of another object.

event horizon Coinciding with the Schwarzschild radius, this is an imaginary boundary surrounding a concentration of mass, such as a collapsing star or black hole. Within the event horizon, no information of the events occurring there can be communicated to the outside, as at this distance not even the speed of light provides sufficient velocity to escape.

extrasolar planet A planet orbiting a star other than the Sun. Over 100 such planets have been discovered, though mostly in highly elliptical orbits.

flat universe This universe results if its density is precisely at the critical level. It is flat in the sense that its space is defined by the rules of ordinary Euclidean geometry—parallel lines never cross.

focal length The distance from a mirror surface to the point where parallel rays of light are focused.

focus The point at which a mirror concentrates parallel rays of light that strike its surface.

frequency The number of wave crests that pass a given point per unit of time. By convention, this is measured in hertz (equivalent to one crest-to-crest cycle per second, named in honor of the nineteenth-century German physicist Heinrich Rudolf Hertz and abbreviated Hz).

galactic bulge Also called nuclear bulge, this is a swelling at the center of spiral galaxies. Bulges consist of old stars and extend out a few thousand light-years from the galactic centers.

galactic disk The thinnest part of a spiral galaxy. The disk surrounds the nuclear bulge and contains a mixture of old and young stars, gas, and dust. In the case of the Milky Way, it extends out some 50,000 light-years from the Galactic center but is only about 1,000 light-years thick. The dust in the disk (a few hundred light-years thick) creates the dark ribbon that runs the length of the Milky Way and limits the view of our own Galaxy (*see* dust lanes).

Galactic halo A large (50,000 light-year radius) sphere of old stars surrounding the galaxy.

galactic nucleus The core of a spiral galaxy. In the case of Seyfert (active) galaxies, the nucleus is extremely luminous in the radio wavelengths.

galaxy cluster A gravitationally bound group of galaxies.

geocentric Earth-centered, the geocentric model of the universe (or solar system) is one in which Earth is believed to be at the center of the universe (or solar system).

giant molecular clouds (GMCs) These huge collections of cold (10 K to 100 K) gas contain many millions of solar masses of molecular hydrogen. These clouds also contain other molecules (such as carbon dioxide) that we can image with radio telescopes. The cores of these clouds are often the sites of the most recent star formation.

gibbous This word from Middle English means "bulging"—an apt description of the Moon's shape between its first and third quarter phases.

globular clusters Collections of a few hundred thousand stars, held together by their mutual gravitational attraction; they are found in highly eccentric orbits above and below the galactic disk.

gnomon Any object designed to project a shadow used as an indicator. The upright part of a sundial is a gnomon.

heliocentric Sun-centered; describes our solar system, in which the planets and other bodies orbit the Sun.

helium flash A stellar explosion produced by the rapid increase in the temperature of a red giant's core driven by the fusion of helium.

homogeneity *See* cosmological principle.

H-R diagram Short for Hertzsprung-Russell diagram. It is a graphical plot of luminosity versus temperature for a group of stars that is used to determine the age of clusters of stars.

Hubble's Law The linear relationship between the velocity of a galaxy's recession and the galaxy's distance from us. Simply stated, the law says that the recessional velocity is directly proportional to the distance and is used to determine the age of the universe.

inflationary epoch A time soon after the Big Bang when the universe was puffed up suddenly, increasing in size by a factor of 10^{50} in an instant. This inflation could account for the incredible sameness, or uniformity, of the universe, even in regions that (without inflation) could never have been in contact with one another.

interferometer This is a combination of telescopes linked together to create the equivalent (in terms of resolution) of a giant telescope. This computing-intensive method greatly increases resolving power.

interstellar matter The material found between stars. Refers to the gas and dust thinly distributed throughout space, the matter from which the stars are formed. About 5 percent of our Galaxy's mass is contained in its gas and dust. The remaining 95 percent is in stars.

interstellar medium *See* interstellar matter.

intrinsic variable *See* variable star.

irregular galaxy A galaxy type lacking obvious structure but containing lots of raw materials for the creation of new stars, and often, many young, hot stars.

isotropy *See* cosmological principle.

jovian planets The gaseous planets in our solar system farthest from the Sun: Jupiter, Saturn, Uranus, and Neptune.

Kelvin scale The Kelvin (K) temperature scale is tied to the Celsius (C) temperature scale and is useful because there are no negative Kelvin temperatures. Absolute zero (0 K) is the coldest temperature that matter can attain. At this temperature, the atoms in matter would stop jiggling around all together. 0 K corresponds to approximately –273°C (–459.4°F).

leading face and **trailing face** Moons that are tidally locked to their parent planet have a leading face and a trailing face; the leading face always faces in the direction of the orbit, the trailing face away from it. Many moons of the outer gas planets are in such locked orbits.

libration The slow oscillation of the Moon (or other satellite, natural or artificial) as it orbits a larger celestial body. Lunar libration gives us glimpses of a very small portion of the far side of our Moon.

light pollution The effect of poorly planned lighting fixtures (such as street and building lighting) that allow light to be directed upward into the sky. Light pollution washes out the contrast between the night sky and stars.

light-year The distance light travels in one year: approximately 5.88 trillion miles (9.46 trillion km). For interstellar measurements, astronomers use the light-year as a basic unit of distance.

(The) Local Group A galaxy cluster, this gravitationally bound group of galaxies includes the Milky Way, Andromeda, and other smaller galaxies.

luminosity The total energy radiated by a star each second. Luminosity is a quality intrinsic to the star; magnitude might or might not be.

lunar eclipse The darkening of the full moon when it passes through the shadow of the Earth. Eclipses can be full, partial, or penumbral.

magnetosphere A zone of electrically charged particles trapped by a planet's magnetic field. The magnetosphere lies far above the planet's atmosphere.

magnitude A system for classifying stars according to apparent brightness (*see* luminosity). The human eye can detect stars with magnitudes from 1 (the brightest) to 6 (the faintest). A first magnitude star is 100 times brighter than a sixth magnitude star. Absolute magnitude is another name for luminosity, but apparent magnitude is the amount of energy emitted by a star and striking some surface or detection device (including our eyes). Apparent magnitude varies with distance.

main sequence When the temperature and luminosity of a large number of stars are plotted, the points tend to fall mostly in a diagonal region across the plot. The main sequence is this well-defined region of the Hertzsprung-Russell diagram (H-R diagram) in which stars spend most of their lifetime. The Sun will be a main sequence star for about 10 billion years.

mantle The layer of a planet beneath its crust and surrounding its core.

maria (pronounced *MAH-ree-uh*) The plural of "mare" (pronounced *MAR-ay*), this word is Latin for "seas." Maria are dark-grayish plains on the lunar surface that resembled bodies of water to early observers.

meteor The term for a bright streak across the night sky—a "shooting star." A *meteoroid* is the object itself, a rocky object that is typically a tiny fragment lost from a comet or an asteroid. A *micrometeoroid* is a very small meteoroid. The few meteoroids that are not consumed in Earth's atmosphere reach the ground as *meteorites.*

meteor shower When Earth's orbit intersects the debris that litters the path of a comet, we see a meteor shower. These happen at regular times during the year.

meteorite *See* meteor.

meteoroid *See* meteor.

micrometeoroid *See* meteor.

millisecond pulsar A neutron star rotating at some 1,000 revolutions per second and emitting energy in extremely rapid pulses.

minor planet *See* asteroid.

nebula A term with several applications in astronomy but used most generally to describe any fuzzy patch seen in the sky. Nebulae (the plural form of the term) are often (though not always) vast clouds of dust and gas.

neutron star The super-dense compact remnant of a massive star, one possible survivor of a supernova explosion. Supported by degenerate neutron pressure, not fusion, it is an entire star with the density of an atomic nucleus.

nova A star that suddenly and very dramatically brightens, resulting from the triggering of nuclear fusion caused by the accretion of material from a binary companion star.

nuclear fission A nuclear reaction in which an atomic unit splits into fragments, thereby releasing energy. In a fission reactor, the split-off fragments collide with other nuclei, causing them to fragment, until a chain reaction is under way.

nuclear fusion This nuclear reaction produces energy by joining atomic nuclei. Although the mass of a nucleus produced by joining two nuclei is less than that of the sum of the original two nuclei, the mass is not lost; rather, it is converted into large amounts of energy.

objective lens In a telescope, the lens that first receives light from the observed object and forms an image.

open universe A universe whose density is below the critical value (*see* critical density). In contrast to a closed universe, an open universe will expand forever because its density is insufficient to halt the expansion.

optical window An atmospheric property that allows visible light to reach us from space.

orbital period The time required for an object to complete one full orbit around another object. The orbital period of the Earth around the Sun, for example, is a fraction over 365 days.

penumbral eclipse When the Moon falls into the lighter, outer part of the Earth's shadow that is not completely dark.

planetary nebula The ejected gaseous envelope of a red giant star. This shell of gas is lit up by the ultraviolet photons that escape from the hot, white dwarf star that remains. (This term is sometimes confusing because it has nothing to do with planets.)

planetary transit An event in which a planet passes in front of the disk of the Sun (or the disk of another star in an extrasolar planetary system). These events in distant systems can be used to probe the properties of the transiting planet.

planetesimals These are embryonic planets in an early formative stage, which are usually the size of small moons, that develop into protoplanets, immature but full-scale planets. Protoplanets go on to develop into mature planets as they cool.

precession The slow change in the direction of the axis of a spinning object (such as the Earth), caused by an external influence or influences (such as the gravitational fields of the Sun and the Moon).

primary mirror In a reflecting telescope, the mirror that first reflects light from the observed object.

primordial synthesis The fusion reactions that occurred in the early universe at temperatures of about 10^{10} K. These fusion reactions produced helium and a small amount of lithium.

proper motion Motion of a star determined by measuring the angular displacement of a target star relative to more distant background stars. Measurements are taken over long periods of time, and the result is an angular velocity (measured, for example, in arcseconds/year). If the distance to the star is known, this angular displacement can be converted into a transverse velocity in km/s (*see* transverse component).

protoplanet *See* planetesimals.

pulsar A rapidly rotating neutron star with a magnetic field oriented such that it sweeps across Earth with a regular period.

pulsating variable *See* variable star.

quasar Short for "quasi-stellar radio source," quasars are bright, distant, tiny objects that produce the luminosity of 100 to 1,000 galaxies within a region the size of a solar system.

radar Short for "radio detection and ranging." In astronomy, radio signals are sometimes used to measure the distance of planets and other objects in the solar system.

radial component *See* transverse component.

radio galaxy A member of an active galaxy subclass of elliptical galaxies. Radio galaxies are characterized by strong radio emissions and, in some cases, narrow jets and wispy lobes of emissions located hundreds of thousands of light-years from the nucleus.

radio jets Narrow beams of ionized material that have been ejected at relativistic velocities from a galaxy's nucleus.

radio lobe The diffuse or wispy radio emissions found at the end of a radio jet. In some radio galaxies, the lobe emission dominates; in others, the jet emission dominates.

radio telescope An instrument, usually a very large dish-type antenna connected to a receiver and recording and/or imaging equipment, it is used to observe radio-wavelength electromagnetic radiation emitted by stars and other celestial objects.

radio window A property of Earth's atmosphere that allows some radio waves from space to reach Earth and that allows some radio waves broadcast from Earth to penetrate the atmosphere.

radioactive decay The natural process whereby a specific atom or isotope is converted into another specific atom or isotope at a constant and known rate. By measuring the relative abundance of parent-and-daughter nuclei in a given sample of material (such as a meteorite), it is possible to determine the age of the sample.

red giant A late stage in the career of stars about as massive as the Sun. More massive stars in their giant phase are referred to as supergiants. The relatively low surface temperature of this stage produces its red color.

redshift An increase in the detected wavelength of electromagnetic radiation emitted by a celestial object as the recessional velocity between it and the observer increases. The name derives from the fact that lengthening the wavelength of visible light tends to redden the light that is observed. By analogy, "redshift" is applied to the lengthening of any electromagnetic wave.

refracting telescope Also called a refractor, this telescope creates its image by refracting (bending) light rays with lenses.

resolving power The ability of a telescope (optical or radio telescope) to render distinct, individual images of objects that are close together.

retrograde An orbit that is backward or contrary to the orbital direction of the other planets.

Schwarzschild radius The radius of an object with a given mass at which the escape velocity equals the speed of light. As a rule of thumb, the Schwarzschild radius of a black hole (in km) is approximately three times its mass in solar masses, so a 5-solar mass black hole has a Schwarzschild radius of about $5 \times 3 = 15$ km.

seeing The degradation of optical telescopic images as a result of atmospheric turbulence. "Good seeing" denotes conditions relatively free from such atmospheric interference.

Seyfert galaxy A type of active galaxy, resembling a spiral galaxy, with strong radio-wavelength emissions and emission lines coming from a small region at its core.

sidereal day A day measured from star rise to star rise. The sidereal day is 3.9 minutes shorter than the solar day.

sidereal month The period of 27.3 days that it takes the Moon to orbit once around Earth. *See also* synodic month.

sidereal year The time it takes Earth to complete one circuit around the Sun with respect to the stars.

singularity The infinitely dense remnant of a massive core collapse.

solar day A day measured from sunup to sunup (or noon to noon, or sunset to sunset) which is slightly longer than a sidereal day. *See* sidereal day.

solar flares Explosive events that occur in or near an active region on the Sun's surface.

solar nebula The vast primordial cloud of gas and dust from which (it has been theorized) the Sun and solar system were formed.

solar wind A continuous stream of radiation and matter that escapes from the Sun. Its effects can be seen in how it blows the tails of a comet approaching the Sun.

spectral lines A system for classifying stars according to their surface temperature as measured by their spectra. The presence or absence of certain spectral lines is used to place stars in a spectral class. *See also* spectroscope.

spectrometer *See* spectroscope.

spectroscope An instrument that passes incoming light through a slit and prism, splitting it into its component colors. A spectrometer is an instrument capable of precisely measuring the spectrum thus produced. Substances produce characteristic spectral lines or emission lines, which act as the "fingerprint" of the substance, enabling identification of it.

spectroscopic binaries *See* visual binaries.

spicules Jets of matter expelled from the Sun's photosphere region into the chromosphere above it.

spiral arms Structures found in spiral galaxies apparently caused by the action of spiral density waves.

spiral density waves Waves of compression that move around the disk of a spiral galaxy. These waves are thought to trigger clouds of gas into collapse, thus forming hot, young stars. It is mostly the ionized gas around these young, massive stars that we observe as spiral arms.

spiral galaxy A galaxy characterized by a distinctive structure consisting of a thin disk surrounding a galactic bulge. The disk is dominated by bright, curved arcs of emissions known as spiral arms. The rotation curves of spiral galaxies indicate the presence of large amounts of dark matter.

standard candle Any object whose luminosity is well known. Its measured brightness can then be used to determine how far away the object is. The brightest standard candles can be seen from the greatest distances.

standard solar model Our current picture of the structure of the Sun, the model seeks to explain the observable properties of the Sun and also to describe properties of its mostly unobservable interior.

stellar occultation An astronomical event that occurs when a planet passes in front of a star, dimming the star's light (as seen from Earth). The exact way in which the light dims can reveal details, for example, in the planet's atmosphere.

summer solstice On or about June 21; this longest day in the Northern Hemisphere marks the beginning of summer.

sunspots These irregularly shaped dark areas on the face of the Sun appear dark because they are cooler than the surrounding material. They are tied to the presence of magnetic fields at the Sun's surface.

supercluster A group of galaxy clusters. The Local Supercluster contains some 10^{15} solar masses.

superluminal motion A term for the apparent "faster than light" motion of blobs of material in some radio jets. This effect results from radio-emitting blobs moving at high velocity toward the observer.

supernova The explosion accompanying the death of a massive star as its core collapses.

synchronous orbit A celestial object is in synchronous orbit when its period of rotation is equal to its average orbital period; the Moon, in synchronous orbit, presents only one face to Earth.

synchrotron radiation Synchrotron radiation arises when charged particles (e.g., electrons) are accelerated by strong magnetic fields. Some of the emissions from radio galaxies are synchrotron.

synodic month The period of 29.5 days that the Moon requires to cycle through its phases, from new moon to new moon. *See also* sidereal month.

telescope A word from Greek roots meaning "far-seeing," optical telescopes are arrangements of lenses and/or mirrors designed to gather visible light efficiently enough to enhance resolution and sensitivity. *See also* radio telescope.

terminator The boundary separating light from dark, the daytime from nighttime hemispheres of the Moon (or other planetary or lunar bodies).

terrestrial planets The planets in our solar system closest to the Sun: Mercury, Venus, Earth, and Mars.

thought experiment A systematic hypothetical or imaginary simulation of reality, used as an alternative to actual experimentation when such experimentation is impractical or impossible.

tidal bulge The deformation of one celestial body caused by the gravitational force of another extended celestial body. The Moon creates an elongation of Earth's oceans—a tidal bulge.

time dilation The apparent slowing of time (as perceived by an outside observer) as an object approaches the event horizon of a black hole or moves at very high velocity.

trailing face *See* leading face.

transit When an object crosses the imaginary half-circle on the sky that runs from north to south (the meridian), it is said to transit.

transverse component Stellar movement across the sky, perpendicular to our line of sight. The radial component is motion toward or away from us. True space motion is calculated by combining the observed transverse and radial components.

triangulation An indirect method of measuring distance derived by geometry or trigonometry using a known baseline and two angles from the baseline to the object.

tropical year A year measured from equinox to equinox. *See also* sidereal year.

universal recession This apparent general movement of all galaxies away from us was first observed by Edwin Hubble. This observation does not mean we are at the center of the expansion. Any observer located anywhere in the universe should see the same redshift.

Van Allen belts Named for their discoverer, American physicist James A. Van Allen, these are vast doughnut-shaped zones of highly energetic, charged particles trapped in the magnetic field of Earth. The zones were discovered in 1958.

variable star A star that periodically changes in brightness. A cataclysmic variable is a star, such as a nova or supernova, that changes in brightness suddenly and dramatically as a result of interaction with a binary companion star, while an intrinsic variable changes brightness because of rapid changes in its diameter. Pulsating variables are intrinsic variables that vary in brightness in a fixed period or span of time and are useful distance indicators.

vernal equinox On this date (usually March 21), day and night are of equal duration because the Sun's apparent course intersects the celestial equator at these times.

visual binaries Binary stars that can be resolved from Earth. Spectroscopic binaries are too distant to be seen as distinct points of light, but they can be observed with a spectroscope. In this case, the presence of a binary system is detected by noting Doppler-shifting spectral lines as the stars orbit one another. If the orbit of one star in a binary system periodically eclipses its partner, it's possible to monitor the variations of light emitted from the system and thereby gather information about orbital motion, mass, and radii. These binaries are called eclipsing binaries.

water hole The span of the radio spectrum from 18 cm to 21 cm, which many researchers believe is the most likely wavelength on which extraterrestrial broadcasts will be made. The name is a little astronomical joke—the hydrogen (H) and hydroxyl (OH) lines are both located in a quiet region of the radio spectrum, a region where there isn't a lot of background noise. Because H and OH add up to H_2O (water), this dip in the spectrum is called the water hole.

wavelength The distance between two adjacent wave crests (high points) or troughs (low points). By convention, this distance is measured in meters or decimal fractions thereof.

white dwarf The remnant core of a red giant after it has lost its outer layers as a planetary nebula. Because fusion has halted, the carbon-oxygen core is supported against further collapse only by the pressure supplied by densely packed electrons.

winter solstice On or about December 21, in the Northern Hemisphere, it's the shortest day and the start of winter.

zonal flow The prevailing east-west wind pattern that is found on Jupiter.

Appendix B

Astronomical Data

Total Solar Eclipses

Date	Duration of Totality (in Minutes)	Where Visible
2008 Aug. 1	2.4	Arctic Ocean, Siberia, China
2009 Jul. 22	6.6	India, China, South Pacific
2010 Jul. 11	5.3	South Pacific
2012 Nov. 13	4.0	Northern Australia, South Pacific
2013 Nov. 3	1.7	Atlantic Ocean, Central Africa
2015 Mar. 20	4.1	North Atlantic, Arctic Ocean
2016 Mar. 9	4.5	Indonesia, Pacific Ocean
2017 Aug. 21	2.0	Pacific Ocean, United States, Atlantic Ocean

Total and Partial Lunar Eclipses

During a total eclipse, the entire lunar surface falls into the Earth's shadow. During a partial eclipse, only a fraction of the lunar surface is deeply shadowed. During a penumbral eclipse, the lunar surface falls only into the less intense "penumbral" shadow of Earth.

Total and Partial Lunar Eclipses*

Date	Type	Duration	Where Visible
2008 Feb. 21	Total	00h51m	Central Pacific, Americas, Europe, Africa
2008 Aug. 16	Partial	03h09m	South America, Europe, Africa, Asia, Australia
2009 Feb. 09	Penumbral	–	East Europe, Asia, Australia, Pacific, west Americas
2009 Jul. 07	Penumbral	–	Australia, Pacific, Americas
2009 Aug. 06	Penumbral	–	Americas, Europe, Africa, west Asia
2009 Dec. 31	Partial	01h02m	Europe, Africa, Asia, Australia
2010 Jun. 26	Partial	02h44m	East Asia, Australia, Pacific, west Americas
2010 Dec. 21	Total	01h13m	East Asia, Australia, Pacific, Americas, Europe

**All eclipses have a penumbral (or less dark) phase, so durations of the eclipse are given only for total and partial lunar eclipses.*

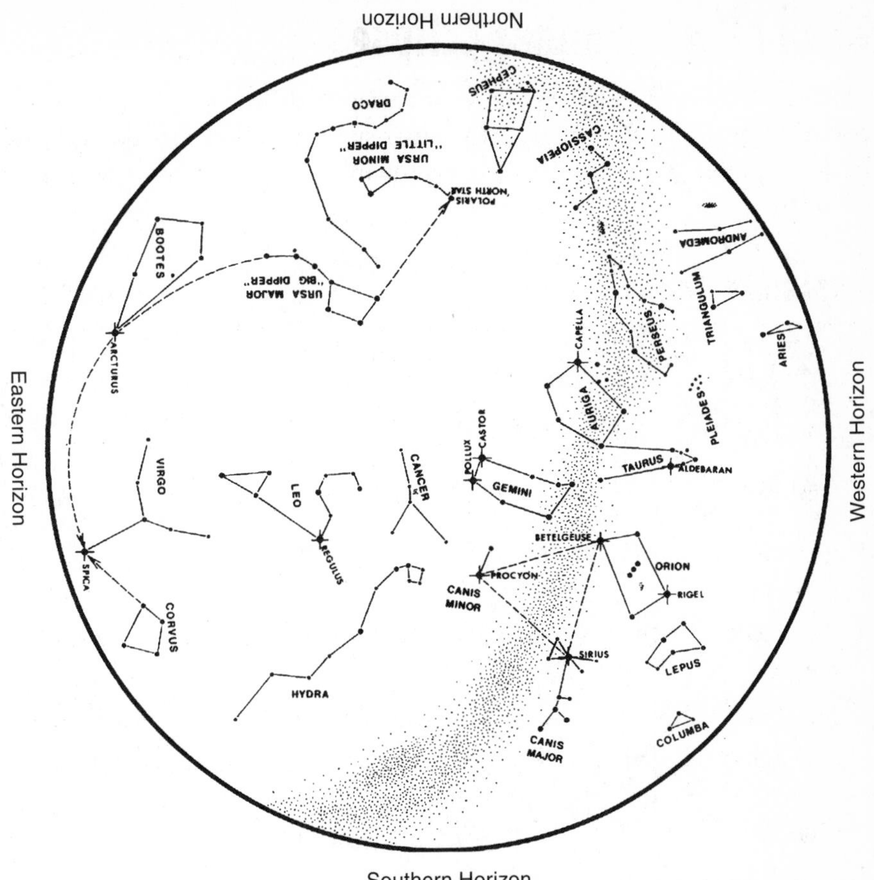

The Night Sky in March

Latitude of chart is 34°N, but it is practical throughout the continental United States.

To use: Hold chart vertically and turn it so the direction you are facing shows at the bottom.

Chart time (Local Standard):

10 p.m. First of month

9 p.m. Middle of month

8 p.m. Last of month

Star Chart from Griffith Observer, *Griffith Observatory, Los Angeles*

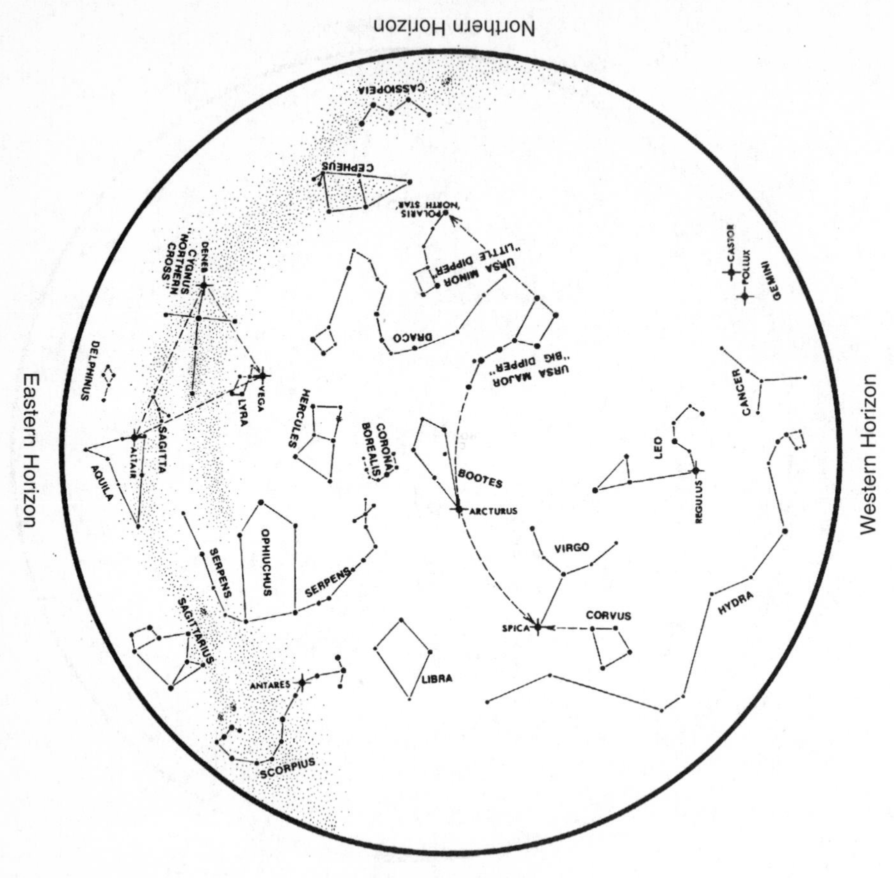

The Night Sky in June

Latitude of chart is 34°N, but it is practical throughout the continental United States.

To use: Hold chart vertically and turn it so the direction you are facing shows at the bottom.

Chart time (Local Standard):

10 p.m. First of month

9 p.m. Middle of month

8 p.m. Last of month

Star Chart from Griffith Observer, *Griffith Observatory, Los Angeles*

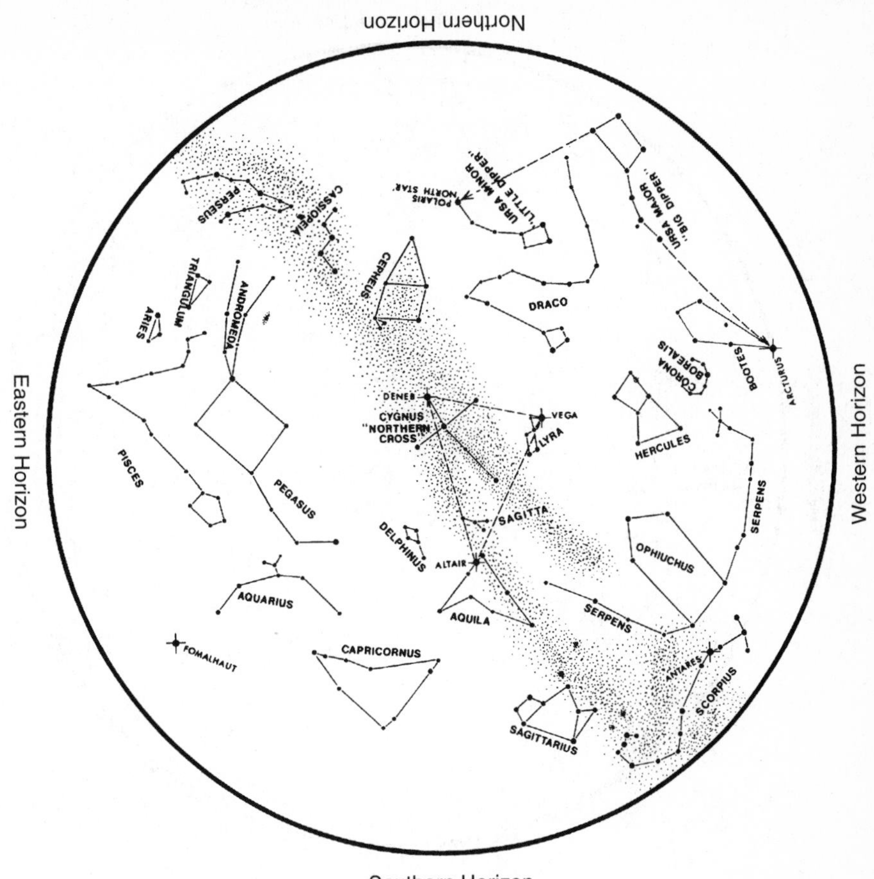

The Night Sky in September

Latitude of chart is 34°N, but it is practical throughout the continental United States.

To use: Hold chart vertically and turn it so the direction you are facing shows at the bottom.

Chart time (Local Standard):

10 p.m. First of month

9 p.m. Middle of month

8 p.m. Last of month

Star Chart from Griffith Observer, *Griffith Observatory, Los Angeles*

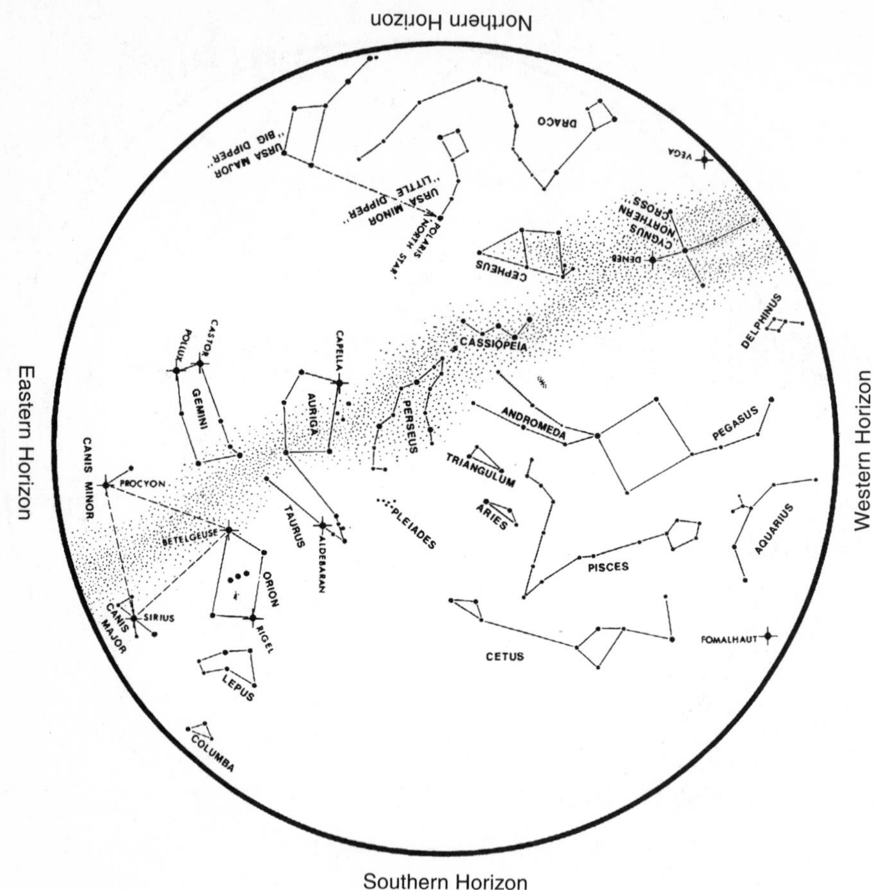

The Night Sky in December

Latitude of chart is 34°N, but it is practical throughout the continental United States.

To use: Hold chart vertically and turn it so the direction you are facing shows at the bottom.

Chart time (Local Standard):

10 p.m. First of month

9 p.m. Middle of month

8 p.m. Last of month

Star Chart from Griffith Observer, *Griffith Observatory, Los Angeles*

Appendix C

Sources for Astronomers

Websites

A wealth of astronomy-related information and images are available on the World Wide Web. We've listed a few highlights for you to check out, but don't hesitate to use a good search engine (such as Google or Yahoo!) to look for additional or more specific information.

Institutions, Magazines, and Societies

Amateur Telescope Makers Association. Valuable for the do-it-yourselfer: **www.atmsite.org**

Astronomy magazine, the site includes links to other astronomy-related sites: **www.astronomy.com/home.asp**

Astrobiology.com. The site is devoted to the maverick science of astrobiology: **www.astrobiology.com**

Institute and Museum of History of Science, Florence, Italy. It includes a wealth of multimedia material (in English) devoted to the work of Galileo: **www.imss.fi.it**

Jet Propulsion Laboratory, a NASA-related facility. You'll find a lot of information on solar system exploration projects and many images. A must-see site: **www.jpl.nasa.gov**

The Planetary Society. Carl Sagan founded this to encourage the search for extraterrestrial life: **www.planetary.org**

SETI Institute. Dedicated to the search for extraterrestrial civilizations: **www.seti.org**

Sky & Telescope magazine. Here's an important source of information (dates, directions) on astronomical events: **www.skytonight.com**

Some Observatories

Anglo-Australian Observatory: **www.aao.gov.au**

Bradley Observatory (on-campus observatory at Agnes Scott College): **www.bradley.agnesscott.edu**

Keck Observatory: **www.keckobservatory.org**

Mount Wilson Observatory: **www.mtwilson.edu**

National Optical Astronomy Observatories: **www.noao.edu**

National Radio Astronomy Observatory: **www.nrao.edu**

Space Telescope Science Institute (source of Hubble Space Telescope images): **www.oposite.stsci.edu**

Space Telescope Science Institute Press Release Page: **www.hubblesite.org/newscenter**

United States Naval Observatory (USNO). Information on timekeeping and sunrise and sunset times: **www.usno.navy.mil**

Yerkes Observatory, University of Chicago: **www.astro.uchicago.edu/yerkes**

Planetary Positions, Solar System

Planetary position calculator: **www.imagiware.com/astro/planets.cgi**

Scale model solar system: **www.exploratorium.edu/ronh/solar_system**

Guides to Events (Eclipses, Meteor Showers, and So On)

Abrams Planetarium Sky Calendar: **www.pa.msu.edu/abrams/SkyCalendar/Index.html**

For solar eclipse information, including link to the SOHO site, with frequently updated solar images at many wavelengths: **www.umbra.nascom.nasa.gov/sdac.html**

Extrasolar Planets

The "go to" website to get current counts of extrasolar planetary systems and recent results: **www.exoplanets.org**

Images and Catalogs

Astronomy Pictures of the Day: **www.antwrp.gsfc.nasa.gov/apod/astropix.html**

The Messier Catalog (a visual catalog of Messier objects): **www.seds.org/messier**

Sky View; truly a virtual observatory: **www.skyview.gsfc.nasa.gov**

Links

The following sites serve as links to many other websites of interest to amateur and professional astronomers:

Astronomical Society of the Pacific. This site includes a link to *Mercury Magazine*: **www.astrosociety.org**

American Astronomical Society (AAS). This site includes links to the online versions of the *Astronomical Journal* and the *Astrophysical Journal*, in addition to information and statistics about the profession: **www.aas.org**

WebStars: Astrophysics in Cyberspace. An impressive list of internet astronomical resource sites and astronomy news: **www.heasarc.gsfc.nasa.gov/docs/www_info/webstars.html**

Magazines

Astronomy (Kalmbach Publishing, PO Box 1612, Waukesha, WI 53187). A popular and well-written monthly journal.

Griffith Observer (Griffith Observatory, 2800 E. Observatory Rd., Los Angeles, CA 90027). Concentrates on the history of astronomy.

Mercury Magazine (390 Ashton Avenue, San Francisco, CA 94112). A publication of the Astronomical Society of the Pacific (ASP); contains historical and scientific articles as well as excellent monthly columns.

Planetary Report (The Planetary Society, 65 N. Catalina Ave., Pasadena, CA 91106). Provides news on exploring the solar system and the search for extraterrestrial life.

Sky & Telescope (PO Box 9111, Belmont, MA 02178). Some consider this the standard for amateur astronomy magazines.

Books

As we told you, you're not alone in your fascination with astronomy. You'll find no shortage of books on the subject. The following are a few that are particularly useful.

Practical Guides

Berry, Richard. *Discover the Stars.* New York: Harmony Books, 1987.

Burnham, Robert, Jr. *Burnham's Celestial Handbook.* New York: Dover, 1978.

Carlson, Shawn, ed. *Amateur Astronomer.* New York: Wiley, 2000.

Charles, Jeffrey R. *Practical Astrophotography.* New York: Springer Verlag, 2000.

Covington, Michael A. *Astrophotography for the Amateur.* New York: Cambridge University Press, 1999.

Harrington, Philip S. *Star Ware: The Amateur Astronomer's Ultimate Guide to Choosing, Buying, and Using Telescopes and Accessories, 2nd ed.* New York: Wiley, 1998.

Levy, David H. *The Sky: A User's Guide.* Cambridge, England: Cambridge University Press, 1991.

Licher, David. *The Universe From Your Backyard.* Milwaukee: Kalmbach Publishing, 1988.

Mayall, R. Newton, et al. *The Sky Observer's Guide: A Handbook for Amateur Astronomers.* New York: Golden Books, 2000.

North, Gerald. *Advanced Amateur Astronomy.* New York: Cambridge University Press, 1997.

Tonkin, Stephen F., ed. *Amateur Telescope Making.* New York: Springer Verlag, 1999.

Webb, Stephen, *Where Is Everybody?* New York: Copernicus Books, 2002.

Guides to Events

Bishop, Roy, ed. *The Observer's Handbook.* Toronto: The Royal Astronomical Society of Canada, annual.

Westfall, John E., ed. *The ALPO Solar System Ephemeris.* San Francisco: Association of Lunar and Planetary Observers, annual.

Star and Lunar Atlases

Cook, Jeremy. *The Hatfield Photographic Lunar Atlas.* New York: Springer Verlag, 1999.

Dickinson, Terence, et al. *Mag 6 Star Atlas.* Barrington, NJ: Edmund Scientific, 1982.

Norton, Arthur P. *Norton's 2000.0, 18th ed.* Cambridge, MA: Sky Publishing Corporation, 1989.

Tirion, Wil. *Sky Atlas 2000.0.* Cambridge, MA: Sky Publishing Corporation and Cambridge University Press, 1981.

Introductory Textbooks and Popular Science Books

Chaisson, Eric, and Steve McMillan. *Astronomy: A Beginner's Guide to the Universe, 2nd ed.* Upper Saddle River, NJ: Prentice Hall, 1998.

Ferris, Timothy. *The Whole Shebang*. New York: Simon & Schuster, 1997.

Fraknoi, Andrew, et al. *Voyages Through the Universe*. Fort Worth, TX: Saunders College Publishing, 1997.

Hartman, William K. *Moons & Planets, 3d ed.* Belmont, CA: Wadsworth Publishers, 1993.

Kaufmann, William J., and Neil F. Comins. *Discovering the Universe, 4th ed.* New York: W. H. Freeman, 1996.

Life in the Universe

Darling, David, *Life Everywhere: The Maverick Science of Astrobiology.* New York, NY: Basic Books, 2001.

Davies, Paul. *The Fifth Miracle: The Search for the Origin of Life.* New York, NY: Simon & Schuster, 1999.

Goldsmith, Donald, and Tobias Owen. *The Search for Life in the Universe, 3rd ed.* Reading, MA: Addison-Wesley, 2001.

Grinspoon, David, *Lonely Planets: The Natural Philosophy of Alien Life.* New York: HarperCollins, 2003.

Ward, Peter, and Donald Brownlee. *Rare Earth: Why Complex Life is Uncommon in the Universe.* New York, NY: Springer, 2003.

Index

A

B

C

D

E

F

G

H

I

J

K

L

M

N

O

P

Q

R

S

T

U

V

W-X-Y-Z